Mindmade Politics

Mindmade Politics

The Cognitive Roots of International Climate Governance

Manjana Milkoreit

The MIT Press
Cambridge, Massachusetts
London, England

This book was set in ITC Stone Sans Std and ITC Stone Serif Std by Toppan Best-set Premedia Limited.

Library of Congress Cataloging-in-Publication Data

Names: Milkoreit, Manjana, author.
Title: Mindmade politics : the cognitive roots of international climate governance / Manjana Milkoreit.
Description: Cambridge, MA : MIT Press, [2017] | Includes bibliographical references and index.
Identifiers: LCCN 2016043621 | ISBN 9780262036306 (hardcover : alk. paper) ISBN 9780262551168 (paperback)
Subjects: LCSH: Climatic change--Government policy. | Climatic change-- International cooperation. | Global environmental change--Government policy. | Cognitive psychology.
Classification: LCC QC903 .M555 2017 | DDC 363.738/74561--dc23 LC record available at https://lccn.loc.gov/2016043621

To all those who will bear the consequences of our collective failure to act timely and decisively.

Contents

Acknowledgments

I would like to express my deepest gratitude and appreciation to Thomas Homer-Dixon, who has inspired and enabled me to work on issues that can make a difference in a world of complex global challenges. This research project would not have been possible with any other mentor. I would also like to thank David Welch for keeping me grounded, sharing his enormous knowledge of international relations scholarship, and offering his time and advice when so many other things were demanding his attention.

It is hard to overestimate the importance of Paul Thagard and his scholarly work for this book. Paul not only developed cognitive-affective mapping, the tool that has enabled me to gather a unique data set for my research and for much of the cognitive theory I draw on, but he has also been an invaluable source of knowledge about cognitive science and the challenges of studying the mind. I thank him for his time and the patient help he brought to our interdisciplinary venture.

Special thanks go to all of my research participants for allowing me to gain a glimpse into their minds. They have been incredibly generous in scheduling and conducting interviews with me and taking part in online Q sorts. Their kind support of my academic research played an essential role in making this book possible.

Over the last three years, I have received the generous support of the Walton Sustainability Solutions Initiatives at Arizona State University's Julie Ann Wrigley Global Institute of Sustainability, which has allowed me to dedicate time and energy to this project. I would like to thank in particular Patricia Reiter for her trust in my ability to build the Imagination and Climate Futures Initiative, which was inspired by the research presented in this book. My thanks also go to Jason Franz and Claire Doddman for their invaluable support of my efforts to master science communication and to

Claire in particular for her extraordinary graphic design skills, as the book cover demonstrates.

Many more people have supported me in the process of writing *Mindmade Politics*. I would like to offer my special thanks to Lucie Edwards, Crystal Ennis, Branka Marijan, Jason Thistlethwaite, and Jennifer Hodbod—amazing colleagues, who provided insights, encouragement, and sometimes just the perspective that I needed to keep going.

Finally, I thank my parents, whose trust, generosity, and limitless support has allowed me to become who I am and to follow my passions, even when they have led me far from home.

List of Abbreviations

ADP	Ad Hoc Working Group on the Durban Platform for Enhanced Action
ALBA	Bolivarian Alliance for the Peoples of Our America
AOSIS	Alliance of Small Island States
BASIC	Brazil, South Africa, India, and China
CAM	cognitive-affective mapping or map
CBA	cost-benefit analysis
COP	Conference of the Parties
CRN	Coalition of Rainforest Nations
EIG	Environmental Integrity Group
EVI	Environmental Vulnerability Index
GAIN	Global Adaptation Index
GEG	global environmental governance
GHG	greenhouse gas or gases
INDCs	intended nationally determined contributions
IPCC	Intergovernmental Panel on Climate Change
LDC	Least Developed Country
LMDC	Like-Minded Developing Countries on Climate Change
NDCs	nationally determined contributions
PA	Paris Agreement
SIDS	Small Island Developing States
UNFCCC	United Nations Framework Convention on Climate Change

1   The Science and Politics of Tears

When we met in 2012, I found Naderev ("Yeb") Saño to be a soft-spoken, sincere, and polite man in suit and tie, every bit the diplomat. Between 2012 and 2014, he was the chief negotiator for the Philippines under the 1992 United Nations Framework Convention on Climate Change (UNFCCC). But, on December 6, 2012, when Saño took the plenary floor at the eighteenth annual Conference of the Parties (COP 18) to the UNFCCC in Doha, Qatar, he did something rather undiplomatic. Breaking down in tears, he declared:

I am making an urgent appeal, not as a negotiator, not as a leader of my delegation, but as a Filipino, I appeal to the whole world, I appeal to leaders from all over the world, to open our eyes to the stark reality that we face. ... The outcome of our work is not about what our political masters want. It is about what is demanded of us by 7 billion people.

It was the day after Typhoon Bopha had hit the Philippines. Almost exactly one year later, he took the same plenary floor in Warsaw at COP 19. This time, he announced: "I will now commence a voluntary fasting for the climate. This means I will voluntarily refrain from eating food during this [Conference of the Parties] until a meaningful outcome is in sight."

Three days earlier, Typhoon Haiyan—one of the strongest storms on record—had hit the Philippines, including Saño's hometown of Tacloban. He talked about the devastation, loss, and suffering caused by the storm. Although he had heard from his brother, he did not know whether the rest of his family had survived. Yeb Saño believed that there was a link between the storm and climate change, but he also believed that "we can fix this. We can stop this madness. Right now. Right here in Warsaw."

If we want to understand why the global politics of climate change are so full of conflict, so frustratingly slow, yet impossible to abandon, we have to pay attention to what Yeb Saño and his colleagues in the global climate change negotiations believe and feel. We need to understand how and why Saño's beliefs are very different from those of many of the negotiators sitting in the plenary sessions he kept addressing. We also need to explore whether his mix of sadness, urgency, and hope can persist another year, and another.

Mindmade Politics argues that a focus on cognition—the elements, structures, and processes of individual and collective thought—is essential for understanding the complex and contentious dynamics of global climate change politics. It makes the case that a cognitive approach can produce novel and relevant insights, identifying potential levers for political change as well as the mechanisms for using these levers. Showing how cognitive processes can be investigated with qualitative methods, the book offers a range of empirical insights into the beliefs of global climate change negotiators.

Continuing disputes over climate change policies at all levels of governance are eroding not only optimism that global solutions can be developed, but also the collective ability of the international community to contain climate change within "nondangerous" limits. The climate change problem presents a moving target for our collective ability to address it (Stocker 2013). As more time passes without effective policies, the potential impacts of climate change are increasing and humanity's abilities to address them—through mitigation, adaptation, or geo-engineering—are diminishing (Rogelj et al. 2013; IPCC 2014). The adoption of the Paris Agreement in 2015 has renewed the optimism of many observers and their trust in the international political process, but, at the same time, the question of effectiveness remains unresolved. The Paris Agreement achieved what was politically feasible, not what was necessary.

This situation raises a fundamental question for a student of international politics. What are the conditions for effective multilateral cooperation on climate change? Much scholarship has been devoted to this issue, offering competing explanations for the absence of cooperation without providing much guidance for overcoming the obstacles identified. Defining climate change as a collective-action problem subject to Garrett Hardin's "tragedy of the commons" (1968), scholars have explained noncooperation with reference to a rationally unwilling hegemon (Falkner 2006; Sunstein 2007); hegemonic rivalry (Paterson 2009) or the lack of a hegemon (Hampson

1989); a regime complex (Keohane and Victor 2011), regime fragmentation (Zelli 2011), or regime ossification (Depledge 2006), or different perceptions of justice across the North–South divide (Roberts and Parks 2006). The key problem from both neorealist and neoliberal institutionalist perspectives is structural: power differences and economic interests favor inaction. Those who have the means to address the problem perceive the costs of climate change mitigation to far outweigh its benefits. From a social constructivist perspective, on the other hand, climate change is an issue of justice, based on ideas of North–South relations, equity, exploitation, and responsibility. Since the structural conditions appear unchangeable on a policy-relevant time scale, and it remains unclear how to untie the Gordian equity knot, scholars of international relations are shifting their attention from the seemingly elusive problem of state cooperation to cooperation by nonstate actors (Schroeder and Lovell 2012), polycentric governance (Ostrom 2010), bottom-up solutions (Betsill and Bulkeley 2006; Pattberg and Stripple 2008; Betsill and Corell 2011; Hoffmann 2011), and alternative governance models such as "minilateral clubs" (Weischer, Morgan, and Patel 2012).

International relations scholarship seems to have reached a theoretical and empirical boundary—one this book seeks to push beyond. It uses recent advances in cognitive science to build a theoretical bridge between major and so far opposing theoretical traditions, thereby deepening readers' understanding of world politics. Along the way, it offers valuable insights into the nature of political thinking, especially in the realm of climate change.

More generally, what I hope to develop here is a type of knowledge that can facilitate change. The necessary starting point for this endeavor is the place where I believe all change starts: in the mind.

A Cognitive Approach to International Relations

Human behavior is purposeful. People act to achieve certain goals, such as protection from harm, generation of wealth, or relief of human suffering. In global climate change negotiations, these goals can include anything from protecting a domestic oil industry to preventing sea-level rise beyond a certain threshold or gaining international support for domestic adaptation challenges. Whatever their content, actor and issue-specific motivations drive any and all observable political behavior.

International relations theories assume that these motivations can be identified and modeled, offering explanations for why and when states go to war, establish international organizations, or disregard a norm. Some of these theories focus on political actors' desires to remain in power, feel secure, or gain a competitive advantage over others. Other theories focus on states' desires to benefit from cooperating with others or to adhere to international norms.

These theories attribute the ability to act purposefully to collective actors, including states, terrorist networks, and transnational NGOs or communities of expert advisors. Such collective, intentional behavior requires shared beliefs about things that exist, things that are desirable, and things that present a threat. It also requires shared beliefs about causality and processes for changing all of these shared beliefs over time. International relations theories have remained largely uninterested, however, in the mental and communicative processes that create these shared beliefs. Instead, they have focused on the behavior of political actors, often making significant and unrealistic assumptions about the motivations behind that behavior.

The central thesis of *Mindmade Politics* is the simple observation that the motivations of political actors, perhaps the most important variables of international relations scholarship, are cognitive variables—products of mental processes.

All political decisions and actions require prior thinking, reasoning, assessing, valuing, and judging, informed by goals, desires, fears, and often complicated sets of motivations. In short, cognition is at the root of all political behavior and decision making. These cognitive processes have been the subject of some of the earliest theorizing in international relations. The most prominent example is the rational-choice framework—the theoretical assumption that actors make decisions as if balancing costs and benefits to maximize the net utility of those decisions. But instead of treating costs and benefits as cognitive entities—thoughts—scholars treated them as material, calculable, tangible realities, such as economic wealth or military power. Goods could be valued and tanks could be counted, they argued, whereas thoughts had none of the characteristics that made them amenable to analysis and comparison.

Theories rooted in rationality and materiality were later countered by another set of theories about human motivations to act. Social constructivism took a keen interest in the intangible part of international life: identities, norms, and the "logic of appropriateness" of behaviors in certain situations. And again, scholars had to turn to tractable variables—verbal and textual expressions of ideas, discourse analysis, process tracing, and interviews. But neither rationalism nor constructivism has developed a serious conceptualization of the processes that generate motivation: human thought, cognition, and emotion.

So far there is no detailed analysis of how individuals (including persons representing states) think about their individual or collective interests in the context of climate change, what they believe to be the applicable norms, and what they perceive as requirements for cooperation. It is unclear whether state representatives in fact have rational thoughts about costs and benefits, structural threats and opportunities, normative considerations of historical responsibility and justice, or a combination of these. What costs do they actually consider? What do they mean by "equity"? And if different individuals hold fundamentally different beliefs, how do they differ and why?

This is a lacuna—both empirical and theoretical—that this book seeks to fill. Identifying what political actors believe today, it offers cognitive explanations for international cooperation or noncooperation on climate change that extend beyond the theoretical explanations offered by international relations scholarship so far. It seeks to answer a central question: what cognitive elements and processes promote or inhibit cooperation to achieve effective responses to climate change?

The subjects of inquiry are the thoughts, beliefs, and emotions of individuals and groups, not the sources of these beliefs (material system structure, scientific information), their carriers (human beings and their political roles) or their consequences (decisions and behaviors). This focus on the mental mechanisms connecting decision-relevant factors and the observed political behavior differentiates my work not only from previous research on climate change politics but also from international relations scholarship more generally. Focused on the mind, a cognitive approach is naturally concerned with issues of agency, intention formation, identity, and the links between thoughts, actions, and political outcomes.

Given my interest in the conceptual content of beliefs, it is important to understand how the cognitive approach outlined here differs in particular from social constructivism. Countering rational-structural theories, social constructivist theories accord causal power to various kinds of ideas, mainly identities and norms, in the belief that ideas themselves produce or at least influence political outcomes. The distinct causal assumption of a cognitive approach is that agents rather than ideas produce political outcomes. An agent is motivated by a specific belief system that provides both the foundation and constraints for political decision making. Further, ideas and beliefs can be understood as the results of mental processes rooted in the biological functions of the human body. Cognition is therefore best understood as a set of observable actor-level processes that bridge the material-social world and its given ideational structure (cognitive input), on the one hand, and political decisions and behavior (cognitive output), on the other. A gate might serve as a useful metaphor: cognition can be conceptualized as the gate through which information about and perceptions of the material and social environment pass in order to lead to a decision or behavior.

Given this gate-keeping function, cognition therefore depends on system structure and the availability of ideas in social discourse. The human mind is embedded in and constantly interacts with multiple social networks, institutional settings, cultural-normative contexts, and the material and natural world. The mind uses all of the signals and inputs it receives from its environment to make sense of the world and enable purposeful action. It responds to these inputs in fairly regular ways—there are stable patterns of thinking and reasoning. But rather than working like a machine with predictable outputs based on a certain set of inputs, cognition can create novelty and surprise. People can change their minds over time, and the cognitive interactions of multiple individuals can produce different and sometimes unexpected results. Small differences in the setup of the "mental machinery" within an individual mind or in the context of a decision can lead to very different outcomes.

The cognitive approach is not a rival perspective to those of international relations' "big three" schools of thought—neorealism, neoliberal institutionalism, and social constructivism—but rather a complementary perspective that is able to speak to present scholarship and potentially integrate past insights across different theoretical schools. Identifying parts of the cognitive status quo in global climate change politics—the cognitive system structure—this research might also provide the foundation for interventions and change.

Mindmade Politics is heavily shaped by three conceptions of cognition that are not without contention. First, adopting the dominant view among cognitive scientists, I reject mind-brain duality and conceive of mental processes as brain processes, rooted in neural activity, chemistry, and, more generally, biological functions of the human body. Second, rather than viewing cognition and emotion as separate systems, I argue that these are intimately connected, indeed integrated, brain functions. Although the focus on the role of emotions in human thought and choice has a long history (e.g., Hume's sentimentalism), it has so far figured very little in international relations scholarship. One of the reasons for this neglect might have been the lack of tools and methods to analyze emotional phenomena. This book makes some progress on both fronts, acknowledging emotions in its theoretical framework and experimenting with new methods to make emotions empirically tractable. Third, I argue that cognition is best understood in complex-system terms. Rather than looking at individual cognitive elements, theories of cognition need to address the relationship between multiple cognitive elements and larger-scale system dynamics that can create worldviews, complex emotions, and strategic political behavior. Sets of cognitive elements and their links can be described as networks of meaning, such as political ideologies (Jost, Nosek, and Gosling 2008), issue frames (Nelson 1997; Benford and Snow 2000), problem-specific or conflict narratives (Smith 2007), or social discourses (Hajer 1996; Maguire 2004). *Mindmade Politics* is concerned with specific networks of meaning: the belief systems of global climate change negotiators with regard to climate change and multilateral cooperation.

Several scholarly developments outside the field of international relations support and contextualize the rationale for focusing on cognition:

1. A growing number of scholars in multiple social science disciplines (e.g., social psychology, communication and decision sciences) are turning their attention to the analysis of cognitive barriers to engagement with climate change at the level of citizens, ideological groups, or local communities. This research is concerned with the role of emotions (Lorenzoni et al. 2006; Wolf and Moser 2011; Roeser 2012), cultural elements (Kahan, Jenkins-Smith, and Braman 2011; Leiserowitz

in Moser and Dilling 2007, chap. 2), ideologies (Weber 2010; Antilla 2005; McCright and Dunlap 2000), communicative strategies including the use of imagery (O'Neill et al. 2013), and the physical experience of climate change–related weather events (Dessai et al. 2004; Myers et al. 2013; Reser, Bradley, and Ellul 2014) as factors in shaping individual responses to and public opinion on climate change (Norgaard 2006a, 2006b, 2011). Shifting from lack-of-information and lack-of-concern explanations to more complex processes in the human mind, these scholars are providing important insights regarding the cognitive barriers to bottom-up political mobilization for action on climate change. Their insights also have important implications for domestic political processes and can be used to improve national and subnational policy making, but their relevance for understanding the UN negotiation process is limited to indirect effects, such as the impact of climate change skepticism in the United States on the international climate science community and the functioning of the Intergovernmental Panel on Climate Change (IPCC) or on congressional politics and the corresponding negotiation strategies of the Obama administration in the UNFCCC process. To have a chance of turning the Paris Agreement into national law without involving Congress, the US delegation at COP 21 worked hard to avoid treaty language that would create new legal obligations for the United States.

2.   There is an increasing academic interest in the history of ideas (Heymann 2010; Jaeger and Jaeger 2010; Weart 2010), as well as the role of different forms of knowledge (Lahsen 2010) and of imagination (Yusoff and Gabrys 2011) in the context of climate change. The underlying assumption of these studies is that ideas—rather than structures or economic power—can shape governance institutions, social structures, and individual lives. More important, when ideas change, the institutions built around them also change. Key issues when exploring the role of ideas for global climate governance have included the goal of "prevent[ing] dangerous anthropogenic interference with the climate system" (Art. 2, UNFCCC; see also Lowe and Lorenzoni 2007; Lenton 2011a) and the political agreement formalized at the 2009 Copenhagen Summit of keeping the global average temperature increase below 2°C (Randalls 2010; Jaeger and Jaeger 2010; Bellamy and Hulme 2011). These studies raise important questions

regarding political actors' motivations and their abilities to use different ideas in their efforts to create a climate governance regime.

In the lead-up to COP 21, the international struggle over ideas focused on the contentious concept of decarbonization—moving toward carbon-free economies—which did not make it into the final agreement. Parties considered alternative terms like "carbon neutrality," "climate neutrality," and "GHG emission neutrality," each with a subtly different meaning and possibly not so subtly different implications for the climate regime. The resulting compromise formulation in the agreement is as far from decarbonization as one could imagine: "to achieve a balance between anthropogenic emissions by sources and removals by sinks of greenhouse gases." Rather than moving toward zero carbon, the idea of the balance implies that carbon can be emitted into the atmosphere as long as there are enough engineering solutions to pull it back out.

Other novel ideas representing similar compromises in the Paris Agreement include "climate resilience" and "low GHG emissions development." The meanings attached to these ideas over the coming decades will have a real bearing on the practice of development and climate governance on the ground.

3. Over the last two decades, there have been notable advances in the cognitive sciences, including the neurosciences, in understanding human thinking and the inextricable link between cognition and emotion (Damasio 1995; Thagard 2006; Moser in Moser and Dilling 2007, chap. 3). This has been accompanied by the development of computational tools for the study of cognition, such as agent-based models (Lustick and Miodownik 2009), neural network models (Thagard 2006), to include artificial neural network models of learning processes (Lindkvist and Norberg 2014), automated text analysis, and cognitive-affective mapping (Findlay and Thagard 2012). So far, however, with a few notable exceptions (Mercer 2005; Mercer 2010; Moïsi 2009; Sasley 2011; McDoom 2012), these novel tools have not been used in the analysis of global politics. And, more important, there have been no studies using them to explore the role of cognition and emotion in multilateral climate change negotiations.

4. Climate change has given rise to a small body of research on the nature of "wicked" or "superwicked" problems, which appear to be intractable.

This scholarship raises questions of how to address these special problems, which presumably are not open to standard political responses (Verweij et al. 2006; Levin et al. 2009, 2012; Prins et al. 2010). The special characteristics of "wickedness" and many issues regarding the appropriate social responses to them remain matters of contention. Do political actors recognize the special characteristics of these problems? How do they acquire relevant knowledge and how does this knowledge affect their beliefs about governance and cooperation?

Together, these developments suggest that focusing on the role of cognition in the analysis of international climate change politics is a promising avenue for generating new and relevant insights. Investigating cognitive processes of individuals and groups that take part in international political processes—diplomats, NGO and private sector representatives—raises important questions regarding the nature, content, and specific characteristics of their thought patterns in contrast to the cognitive responses of citizens and domestic political actors.

Critics might argue that a cognitive approach is too reductionist, claiming that all political phenomena can be traced back to individual psychology. The previous paragraphs have already begun to rise to the reductionist challenge, outlining a definition of cognition that does not seek to reduce all political phenomena to the processes of an individual mind. Instead, this approach conceptualizes the mind as embedded in material and social environments, which heavily shape all mental processes. Instead of being reductionist, this approach is systemic, viewing cognitive processes as components of larger social-material systems, in which minds have important mediating functions between system conditions and political behavior.

Rather than seeking to explain political phenomena with individual thought processes, the cognitive approach presented here places emphasis on the relationship between individual and collective beliefs, what I call the "person-group problem." The ontological status of collective cognition presents a fascinating theoretical puzzle with relevance across several social science disciplines. For example, much international relations scholarship treats the state not only as a unitary actor, but also often as a thinking and feeling unitary actor, which assumes that a collective can have mental processes. Similarly, economists assume that businesses and other organizations are unitary actors with specific intentions and the ability to make decisions ("Google was concerned about ..."). And sociologists attribute mental capacities to social groups and whole societies. For its part, however, cognitive science makes no such sweeping and simplifying assumptions; it offers the necessary conceptual tools to grapple with the relationship between individual and collective cognition in the absence of a collective mind.

Even so, one can still question the value of a focus on individual minds because such an approach fails to acknowledge the relevance of all social processes, conditions, and institutions to the creation of individual beliefs. If one assumes, as Mary Douglas (1986) does, that the social is prior to individual cognitive processes and in fact shapes and determines individual thought, then a focus on individual minds would seem to confuse independent and dependent variables.

That said, neither the individual nor the social deserves to be prioritized. Rather than focusing on one or the other, *Mindmade Politics* contends that the interaction and mutual interdependence between the individual mind and the social environment are key to understanding existing beliefs and belief dynamics. It is unclear how much cognitive "freedom" or self-determination rests with the individual, and to what extent a person's beliefs have been received from the social environment or even imposed by it. But even when acknowledging the crucial interactions between the individual and the social, much can be gained from using the individual mind as an entry point to the analysis. Most important, the rules that apply to the individual brain condition what types of belief systems are possible, and those constraints also apply to shared or collective beliefs. Unless a belief system first exists in an individual mind, it cannot be shared among members of a group. Further, understanding individual cognitive processes allows researchers to understand how social processes affect individual beliefs—which ideas are adopted or rejected and why. A focus on social factors cannot explain how these social phenomena came into existence in the first place—every institution, practice, or ideology had to start with a set of ideas and arguments in an individual mind. Finally, individual minds are also a source of novelty and disruption—an indispensable source of change in existing social structures. Consequently, a focus on cognition, defined as individual mental processes, offers potentially valuable insights that cannot be gained with a focus on the social.

This explorative study presents initial empirical insights concerning current cognitive patterns among political actors engaged in multilateral negotiations with the aim of advancing theories about the international politics of climate change, the science-policy interface, and multilateral cooperation. The main research question—what cognitive elements and processes promote or inhibit cooperation to achieve effective responses to climate change?—raises two distinct subquestions. What are the most important and perhaps most common cognitive elements and processes in the minds of climate change negotiators? And how do these elements and processes impact ongoing political efforts to create a cooperative multilateral agreement on climate change?

I assume that it is possible to identify specific types of cognitive elements (e.g., concepts, beliefs) or processes (e.g., risk assessments) that are recurrent in the belief systems of different individuals, regardless of their specific views on global climate governance. I distinguish two types of cognitive elements. First, there are general cognitive elements that are essential for an actor's thinking about any issue in international politics. These are the fundamentals of political thought without which an actor would not be able to function or make sense of reality. These elements include, for example, an actor's self-representation (i.e., concepts related to the concept "my country" or "my state"), concepts regarding other actors (e.g., the nature and attributes of other states, NGOs, international organizations, etc.), and concepts regarding the relevant structural, institutional, and normative context. Within this broad category of general cognitive elements, I focus on three subsets of concepts: those related to (1) rational choice (i.e., costs and benefits); (2) self-identity ; and (3) norms of justice. These three sets of concepts reflect well-known theoretical categories in international relations scholarship and the building blocks of political life.

Second, there are cognitive elements that are specific to climate change. These are mental representations of the special characteristics of the climate change problem, such as climate tipping points, greenhouse gas (GHG) emissions, or temperature targets. These concepts heavily shape contentions about climate governance, from appropriate long-term goals to the technicalities of monitoring treaty compliance. In exploring them, I was particularly interested in three characteristics of climate change and negotiators' cognitive responses to them: (1) the overwhelming complexity of the problem, potentially creating a sense of hopelessness; (2) uncertainty and the long time horizons of climate change, requiring long-term thinking in a myopic political context; and (3) the imperceptibility of climate change for the average person (and negotiator) on a daily basis, contributing to a lack of urgency.

With this distinction clearly in mind, I wondered whether these two types of cognitive elements—especially the interaction between general political concepts and the specific concepts related to special characteristics of climate change—have any discernible effect on actors' ability to develop cooperative responses to climate change. In other words, what types of concepts and thought processes are relevant for cooperative or noncooperative decisions and behaviors in global climate change politics?

I approached these questions in three steps. Step one aimed at identifying relevant concepts and concept categories for individual and collective decision making. Step two consisted of a qualitative analysis of the ways in which study participants' minds deal with the special characteristics of climate change. Step three brought these two pieces together, analyzing their interactive effect on the ability of international policy makers to agree on effective, cooperative solutions to climate change. Step one is about cognition in international decision making generally—what are the relevant concepts? Step two is issue specific, asking whether the general cognitive features are influenced by the nature of the problem at hand. Step three explores the relevance of these basic insights for the chances of international cooperation.

Methodology: New and Established Tools to Explore Subjectivity

In contrast to its more conventional counterparts, my research design identifies, visualizes, and analyzes subjectivity rather than material variables or behavior. It deploys two complementary tools—cognitive-affective mapping and the Q method—to identify the content and structure of participants' belief systems. The empirical work (interviews and Q sorts) was conducted in the spring and summer of 2012. Each data source offered a quite different, but complementary perspective into the minds of the study participants. Based on a semistructured interview, a cognitive-affective map (CAM) provided a "cognitive snapshot," insights into the views of a specific individual at a particular point in time. A CAM is a network diagram or concept graph that "displays not only the conceptual structure of people's views, but also their emotional nature, showing the positive and negative values attached to concepts and goals" (Thagard 2012, 37).[1] The Q method identifies different ideal-type belief systems (factors) that are shared by a number of individuals. Cognitive-affective mapping, the Q method, and the complementary strengths of these two tools will be described in greater detail in chapter 4.

The Central Arguments

At its core, *Mindmade Politics* argues that we cannot fully understand political behavior without taking into consideration the thoughts, complex motivations, and emotions of political actors. This is not to say that cognition is all that matters—far from it. Other factors, such as institutional processes or material power shape political outcomes and, indeed, often offer a simpler explanation for those outcomes. The strength of current international relations scholarship lies in categorizing and isolating these different drivers of political behavior. A cognitive approach, for its part, provides a glimpse into the minds of decision makers; it lets us see which of those drivers actually mattered in the political process. Did a state use force to obtain a material payoff or to enforce an international norm? Did two countries form an alliance to balance the power of a third or to create cooperative benefits for each other? What do states expect when they establish a new institution? Are they sometimes compelled by an international norm rather than the expectation of material gain? Using special tools to answer these questions empirically, a cognitive approach not only provides a new kind of data source for social constructivism; it also provides insights into the kinds of rational reasoning that take place among global political actors. What costs and benefits do these decision makers take into consideration? How do they treat nonquantifiable costs? Which goals do they value more than others? As an added benefit, the cognitive approach reveals how rational and constructivist variables are linked and interact in decision makers' minds. Breaking down long-standing instances of entrenched resistance to change ("silos") in the discipline and bridging theoretical perspectives often seen as mutually

---

1. The process for generating and analyzing CAMs is described in detail in "Process: Generating CAMs" and "CAM Content Analysis" in chapter 4.

exclusive, these linkages raise new, interesting questions and offer fruitful avenues for future research.

The following chapter sections briefly sketch the key arguments of the book, focusing on insights that benefited from this interdisciplinary approach.

The Cognitive Triangle: Cost, Identity, and Justice

One of the most useful findings of this book's cognitive analysis is that actors in global climate change politics are motivated by rational reasoning, norms, and ideas at the same time. What is more, the links between rationality and normativity follow a pattern that is likely to apply in other political contexts as well.

More concretely, the link consists of a cognitive interdependence of three variables that are usually treated separately in international relations scholarship: rational cost or threat assessments, collective identities, and normative-ethical considerations. The cognitive counterparts to these well-known variables relate to, influence, and condition one another, jointly constraining the types of arguments different climate change negotiators make publicly. I call this interdependent relationship the "cognitive triangle."

The central elements of all belief systems identified in this study are collective actor identities. All other elements of a person's belief system, including the nature of the problem at hand, are defined and understood in relation to a specific in-group the individual identifies with. Climate change as an environmental phenomenon, the types of risks it poses, the actions necessary to address it, and the moral norms associated with efforts to solve the problem all depend on an actor's vantage point and self-identity. Put differently, self-identity conditions an actor's rationality by shaping—enabling and constraining—perceptions of costs and benefits in a given system structure. This insight validates both structural and constructivist theories, but it also emphasizes that a full understanding of political behavior is not possible without reference to both theoretical schools.

Depending on the constellation of identity conceptions, risk perceptions, and associated emotions, individual actors adopt very different moral frameworks. Two fundamentally different types of belief system can be distinguished. In the first, the actors believe that their in-group faces existential threats, identity loss, death, or grave human suffering. These concerns trigger strong negative emotions and are associated with a deontological framework of reasoning that infuses the actors' negotiation positions with categories of right and wrong. International relations scholars would call this a belief system with a dominant logic of appropriateness. Given the nature of climate change and its differentiated global impacts, the likelihood of having such a deontological belief system increases whenever actors identify with groups larger than the state, for example, the group of developing countries, the poor, or even humanity.

In the second type of belief system, actors focus on material risk types related to climate change rather than survival and suffering. These material risks include economic loss or infrastructure damage and tend to be associated with climate change policy rather than direct climate change impacts. Although such costs are undesirable, they do not trigger the same kind of strong, negative emotions that are associated with concerns about survival and human suffering. Absent these strong emotions, the dominant moral framework in this second type of belief system is consequentialist. Since consequentialism strongly aligns with the results of the actor-specific rational calculation of costs and benefits, the logic of consequences dominates this mental framework.

They key to understanding these differences in beliefs are specific self-identity concepts and risk perceptions of individual actors who see themselves as members of different groups, ranging from local communities to the human community. Urgency and support for action on climate change exist if actors perceive a particularly grave type of threat to their respective in-groups within a relevant time frame. If, however, the actors perceive that the threat is of a less severe type, that it does not affect their in-groups, or that it is too distant in the future, the link between urgency and support for action is broken. When applied to the climate change negotiations with the current distribution of climate change impacts and the current distribution of beliefs about future impacts, this observation explains why some negotiators are much more supportive of ambitious, immediate collective action than others. Representatives of small-island states need only to have a strong sense of connectedness with their fellow citizens to feel urgency and a desire to cooperate globally. But representatives of wealthy Western democracies need to have a more cosmopolitan self-identity—they must feel themselves connected to all humans—for the same cognitive experience.

In essence, a cognitive analysis reveals a previously hidden bridge between structural-rational and constructivist theories.

The Relevance of Structure

Much research on the climate change negotiation process has pointed to the importance of structural explanations—the unequal distribution of power and vulnerability—for its slow progress and the heavily contested issue of sharing the global mitigation burden. *Mindmade Politics* confirms and offers some nuances to this conventional wisdom. Study participants thought about and perceived the importance of the international system structure in two distinct ways. First, they made a range of distinct arguments concerning the global distribution of power in terms of the economic interests of individual states, the interests of vested domestic interest groups, and the relationship between the developed and the emerging economies. Second, they acknowledged the relevance of cross-scale linkages and the influence of domestic politics in the global climate governance system, especially in the case of the United States.

Science in Climate Change Negotiations

Scientific insights, especially those concerning the special characteristics of climate change, did not play a significant role in shaping the belief systems of study participants concerning global climate governance. And because the participants did not acknowledge, or pay much attention to most special problem characteristics (e.g., climate tipping points), they generally did not use these characteristics to motivate arguments about needed solutions, governance instruments, or goals.

Concepts about time are crucial for climate governance. A phenomenon with exceedingly long time horizons, climate change implies a causal impact of today's decisions deep into the future. The special temporal extent of this phenomenon inevitably gives rise to major, sometimes irreducible, scientific uncertainties, for example, regarding the timing and scale of future climatic changes and their impacts on human societies. Uncertainties also exist concerning the response of the climate system to a range of policy measures from emission reductions to carbon removal, and it is impossible to predict how future societies will respond to a changing climate.

But, based on the observed behavior of my study participants, when faced with the task of thinking about the distant future, the cognitive systems of climate change negotiators exhibit significant shortcomings (lack of proactive imagination, reluctance to acknowledge possibility of governance failure) or even lack of attention (avoidance). Individuals are not able to feel and anticipate an imagined future with the same intensity that accompanies their memories of the past or their experience of the present. Consequently, the distant future is generally underdefined, underrepresented, and undervalued in the negotiations.

Agency and Hope

Drawing on these different observations about the beliefs of study participants with regard to climate change negotiations, I discuss their implications for actors' perceptions of their own agency in the global climate governance process. Contrasting two distinct definitions of agency, one rooted in international relations and one in psychology, I argue that political theories of agency would benefit from two conceptual expansions. First, insights from psychology and this study suggest that the inclusion of time—concepts and emotions associated with the past and the future—affect agents' decisions and behavior in the present in multiple, so far poorly understood, ways. Second, hope could be a relevant dimension of agency that has not yet been fully explored either in international relations or psychology.

The absence of clear timelines for climate change impacts, policy actions, or expected responses of the climate system to these actions has important implications for selecting appropriate climate governance goals and targets. The 2°C temperature target adopted at the 2009 Copenhagen Summit was not associated with a clear timeline, specific actions, or milestones. Not able to imagine how the target could be reached, and observing the slow political progress in the UNFCCC as of 2012, many negotiators were pessimistic about their collective ability to reach this target. They had begun to mentally abandon the temperature goal, replacing it with a goal that was more certain and within their collective skill set: a political agreement to be reached by the end of 2015. Now that this diplomatic goal is accomplished, the temperature target has moved back into the spotlight again, but is no more realistic or better understood than it was in 2012. Surprisingly, the 2015 Paris negotiations reintroduced the more ambitious goal of keeping average global warming to less than 1.5°C above preindustrial levels—a troubling proposal I will explore in greater depth in chapter 4. But the final agreement text removed all references to a specific timeline. The inconsistency between the long-term target and the rest of the Paris Agreement is an indicator for a larger psychological problem related to global goal setting. Temperature targets mean very little to climate change negotiators and do not have the motivating force a good goal is supposed to unleash. If these goals are symbolic, rather than driving actual regime performance and effectiveness, they serve no useful function. Even worse, if they are used as bargaining chips rather than serious outcome states, they can create rather than solve political problems. All of this raises the question what would be an appropriate and effective climate regime target?

In addition to discussing these specific cognitive patterns and their effects on global climate change politics, *Mindmade Politics* outlines and discusses six private belief systems that were prevalent among study participant negotiators in the Umbrella Group, the European Union (EU), some members of the G77 and China, and a broad range of NGO representatives. I have labeled these belief systems "Multilateralism champions," "UN skeptics," "Utilizing the market," "The power of individuals," "Climate justice," and "Spotlight on the West." These six systems share a number of fundamental ideas but differ in their views on climate governance. Each focuses on the responsibilities of a different actor group, ranging from individual states, to groups of states to individual human beings. One can distinguish belief systems that are dominated by norms (e.g., international solidarity, the rich help the poor) and those dominated by a logic of costs or consequences, while both logics always coexist.

My research project has two important limitations. First, it deals with the international politics of climate change, not with global affairs generally or domestic climate change politics specifically. Owing to the project's focus on the special problem characteristics of climate change, the theoretical framework I have developed has only limited applicability to other global governance challenges or international relations more generally. Focusing, as it does, on global climate change politics, my project investigates the cognitive processes of a particular group—those engaged in global climate change negotiations—rather than people in general, citizens of a particular country, or other social groups. This particular group consists of several subgroups, including state representatives (diplomats/negotiators), representatives of NGOs, and representatives of corporate actors (firms, business associations).

And, second, my project deals with cognitive content—the concepts, belief systems, and patterns of reasoning that are specific to climate change—not with the general political psychology of international climate change decision making, which concerns various cognitive mechanisms that bound rationality. Psychology and behavioral sciences explore these mechanisms extensively, for example, prospect theory (Kahneman and Tversky 1979), anchoring, or the availability heuristic (Slovic 2000).

A Guide to the Book

Mindmade Politics is primarily written for researchers, educators, and students interested either in climate change politics or in novel applications of cognitive theory. Each of these two audiences—readers with a background in either political science or cognitive theory—will find a brief introduction to the other's discipline in chapters 2 and 3. Practitioners of climate change politics—diplomats, policy advisors, and representatives of NGOs or the corporate world—will find the empirical chapters (5 and 6) most interesting and might want to skip much of the theoretical and methodological discussions in the first half of the book. Chapter 7 offers practical advice for those readers, suggesting ways in which the insights of this book might inform their work.

Chapter 2 presents a short survey of the scholarly literature on the global politics of climate change, its challenges, and its most promising current directions. Identifying a number of lacunas that a cognitive approach begins to fill, the chapter also touches on a range of relevant literatures in other disciplines, including psychology and the decision sciences, that provide important scholarly context and input for a cognitive analysis of global climate change politics.

Chapter 3 introduces the cognitive concepts that inform this research, setting up the conceptual foundations and key components of a cognitive analytic framework for global climate change politics. The chapter's core concern is the definition of cognition for the purpose of analyzing this specific political area. The chapter probes cognitive assumptions of major international relations theories and integrates them into a preliminary conceptual framework that guides the remainder of the book.

Chapter 4 dives into matters of research design, including participant selection, methods, and tools. Zeroing in on the challenging question how qualitative cognitive research can be operationalized, it outlines the rationale for the design of this study and the synergies between the tools I have chosen to work with.

Chapters 5 and 6 present the central empirical arguments of *Mindmade Politics*, based on cognitive-affective mapping and the Q method, respectively. The insights outlined in chapter 5 are the result of a comparative analysis of the belief systems of fifty-five individual study participants engaged in the global climate change negotiations, using the visual and analytical support of cognitive-affective maps. The chapter develops four major themes in response to the central research question about the role of cognition in global climate change politics. First, it argues that there is a cognitive interdependence between three variables that are usually treated separately in international relations scholarship: rational cost (threat) assessments, collective identities and justice concerns. Second, it outlines the role of the system structure for climate change politics—what international relations scholars call the third image. Third, it explores whether and how climate change negotiators make use of scientific information to form political beliefs, and to what extent the special characteristics of climate change play a role. And, fourth, the chapter discusses the relationship between hope and agency—one of the key concepts in political science. It makes the case for a conceptual expansion of current theories of agency in political science, accounting for the role of time and emotions.

Chapter 6 offers a description and comparison of six different belief systems of study participants engaged in the UNFCCC negotiations, each centered on a specific type of actor (e.g., state, market, or individual) and each with its own cognitive-emotional logic. The analysis highlights a set of beliefs that was shared among all participants and formed the minimum consensus and motivation for continuing negotiations. It also identifies a number of highly contentious ideas, and offers a preliminary assessment of the role of emotions in these different belief systems.

Using key insights of the book to develop a number of recommendations for diplomats, policy experts, activists, science communicators, and other political actors taking part in global climate governance, chapter 7 offers a guide for practitioners, with suggestions for reflective-analytic habits and exercises, as well as communication strategies and interaction tools that could be used in the negotiation process. The chapter also contains a call to action for science communicators and knowledge brokers. Using a number of challenges identified throughout the book, I hope to start a conversation about novel ways to bridge the many gaps that currently exist between scientific knowledge and political meaning, understanding and action.

The conclusion, chapter 8, ties together the findings of previous chapters and emphasizes their relevance for the scholarly and policy-making communities. Offering thoughts on future directions of a cognitive research program within political science, I make the case for an ambitious effort to build a cognitive theory of international relations that could bridge major strands of existing theory without abandoning their rich insights.

The empirical data for *Mindmade Politics* were gathered in 2012, at a midpoint between the failed Copenhagen Summit in 2009 and the adoption of the Paris Agreement in 2015. The Durban Conference of the Parties (COP 17) in 2011 had just begun the work of the Ad Hoc Working Group on the Durban Platform for Enhanced Action (ADP), and the parties were entering a search for novel ideas for global climate governance and novel approaches for addressing deeply contentious political issues. The heavy cloud of the Copenhagen failure still hung over the UNFCCC community, but was starting to disperse. The year 2012 was an important inflection point, a time when both creativity and pragmatism were needed, when the post-Copenhagen pessimism and misgivings had to make room for optimism and engagement. *Mindmade Politics* offers a glimpse into this important moment in the history of climate change politics. It reveals some of the sources of novelty that ended up shaping the Paris Agreement, but also a number of long-standing cognitive features of climate diplomacy and political reasoning more generally. Writing in the aftermath of the Paris Agreement, I reflect on both the stable and the dynamic cognitive features that influence the climate change negotiations, and what they mean for the success of a global climate regime that is still in the making.

2 International Politics of Climate Change

This chapter offers a map through some of the social science landscape that deals with climate change politics at the global scale. It deliberately leaves out all research on domestic climate change policy (e.g., cap and trade vs. carbon taxes, adaptation, renewable energy expansion) and on domestic climate change politics, focusing instead on the inter- and transnational dimensions of climate change politics and policy.

The political science scholarship on climate change and international cooperation with regard to it emerged in the 1980s, lagging a bit behind the actual science of climate change. The current chapter sketches the contours of three distinct fields of research that provide important conceptual foundations for the remainder of the book. It begins with international relations—the study of state interactions and global regime-building efforts to address climate change. Escaping the traditional conceptual constraints and state focus of their discipline, many international relations scholars interested in climate change have joined the interdisciplinary earth system governance (ESG) community or, more generally, the global environmental governance (GEG) community. The chapter touches on some of the ESG themes, such as transnational actor networks and governance experiments, before turning to the relationship between science and global politics.

This brief journey through a vast landscape of scholarly work over the last twenty-five years will help readers understand not only some of the intellectual roots of today's thinking about and major scholarly contributions to global climate change politics, but also the current boundaries of knowledge in that regard, which *Mindmade Politics* seeks to push beyond. Unsurprisingly, many of the themes and questions explored by the research community are also part of the belief systems of climate change negotiators.

That overlap between scholarship and cognitive reality opens up a new set of questions and opportunities for deeper communication at the interface of science and policy.

States, Diplomacy, and Building International Regimes

Climate change is a collective-action problem of the kind that Garrett Hardin (1968) described as subject to the "tragedy of the commons"— namely, that the rational overexploitation of the atmosphere by individual actors (persons or states) results in the undersupply of a global public good (climate stability), or worse, in the collapse of the shared resource. Conventional collective-action theory predicts that cooperative solutions cannot be found unless an external authority (a "hegemon" in political science jargon) establishes a rule-based system and penalizes rule infringement and free riding (Ostrom 2010).

Applying this collective-action logic to the international system, much international relations scholarship is concerned with the question why and under what conditions states cooperate to provide global public goods. This is a particularly interesting question given that the presumed equality of sovereign states implies that there is no external or higher authority (i.e., world government) that could establish and enforce rules. In the absence of such an authority, why would states cooperate with one another—in essence, impose rules upon themselves? The standard answer to that question is "When it is in their interest." Defining "national interest" is a tricky business, though. Most scholars assume that cooperation has to provide some clearly identifiable benefit that outweighs its costs in order to be in the national interest. But actors' perceptions of costs and benefits can vary tremendously. In the case of climate change, the benefits of cooperation have not received nearly as much attention as the costs of action to address the problem. At the same time, debates about the costs of mitigation have largely ignored the costs of climate change itself—impacts more diverse and surprising than any political scientist (or economist for that matter) could have imagined.

If and when states cooperate, this usually takes the form of international regimes—"social institutions that consist of agreed upon principles, norms, rules, decision-making procedures, and programs that govern the interactions of actors in specific issue areas" (Young 1997, 5–6). The climate governance regime currently consists of the UNFCCC and all the institutions, activities, and agreements that have been established since its inception in 1992. This includes international treaties like the Kyoto Protocol (1997) and, most recently, the Paris Agreement (2015), but also less formal political agreements like the Copenhagen Accord (2009). But, to date, this regime and the broader regime complex (Keohane and Victor 2011) remain weak and ineffective. More than twenty years of continuous negotiations have brought about an explosion of institutions (e.g., Green Climate Fund, Adaptation Committee, Climate Technology Center and Network), rules (e.g., reporting, monitoring, and accounting requirements), mechanisms (e.g., Clean Development Mechanism, the Financial Mechanism, the Warsaw International Mechanism for Loss and Damage), and processes, most recently that of having states submit nationally determined contributions (NDCs)—voluntary promises of action—every five years. Nevertheless, global greenhouse gas (GHG) emissions continue to be on the rise, and the rate of climate change seems to be accelerating. In 2014, global emissions of carbon dioxide (CO_2) remained stagnant for the first time since measurements began, but there is little confidence that this marked the peak of GHG emissions, rather than representing a temporary slowdown. In sum, for all the signs of regime building, there are no signs of reining in climate change.

After the failed Copenhagen Summit in 2009, efforts to strengthen the global climate governance regime with a new international agreement were viewed with a mixture of exasperation, hopeful optimism, and experience-fueled doubt. These efforts culminated in Paris in 2015, when the UNFCCC member states adopted a new international treaty—the Paris Agreement, which entered into force on November 4, 2016, much sooner than most negotiators or observers had expected. The agreement has been both praised as a landmark success and criticized as a massive failure or even a fraud. This mixed reception reflects the nature of the international regime—it is necessary to advance, yet unable to succeed. Paris achieved, and could only achieve, what was politically feasible rather than what would have been necessary to slow climate change. Most of all, it laid the foundation for even more regime building in the future.

The scholarship on international climate change politics has mirrored the reality of frustratingly slow negotiation progress over two decades, offering multiple explanations for the failure of states to produce an effective cooperative regime (Michaelowa and Michaelowa 2012; Bernauer 2013). Much of this scholarship is dominated by rational choice–based approaches, seeking rational explanations for the failure to supply a global public good (Ward 1996; Grundig 2006; Keohane and Victor 2011). Cass Sunstein (2007, 5) concluded that for the United States "the monetized benefits of the Kyoto Protocol would be dwarfed by the monetized costs" and consequently the United States was rationally opposed to support the climate governance regime. On the basis of hegemonic stability theory, Robert Falkner (2005) argued that, since domestic support for an effective climate change treaty was lacking, but the costs for exercising its hegemonic leadership to implement such a treaty would be very high, the United States would simply not do so. And the United States has so far confirmed this rational perspective with its negotiating position, consistently refusing to accept a treaty that fails to establish mitigation obligations for the emerging powers, particularly China and India, allowing them to free ride on its costly efforts to curtail climate change. The Paris Agreement finally addressed this US concern, establishing the same obligation for all participating countries, including the emerging powers, to submit recurrent nationally determined contributions (NDCs) toward achieving shared global goals. At the same time, neither the United States nor any other nation made a legal (and possibly costly) commitment to reduce emissions. Paris did away with legally binding obligations in favor of recurrent nonbinding promises to take domestic action. Whether the long-term outcomes of this agreement are in any state's national interest remains to be seen. But they are certainly consistent with a rational-choice framework of international politics.

The rationalist pessimism about international cooperation on climate change has led many scholars to advocate for multilevel or polycentric governance (Rayner 1991; Ostrom 2010), shifting attention away from states to substate and nonstate actors (Pattberg and Stripple 2008; Betsill and Bulkeley 2006; Andonova, Betsill, and Bulkeley 2011). This trend goes hand in hand with a critique of the regime-based approach to international politics. Its focus on the global as a distinct sphere of politics artificially separates the tightly interconnected layers of a multiscale process; it undervalues nonstate actors, uses an incomplete top-down perspective, and fails to address the exercise of power (Bulkeley and Newell 2015, 10–11). Other scholars have been looking for change within the regime framework, for example, exploring desirable changes in the process and structure of the multilateral negotiation setting (Eckersley 2012) or advocating that we simply wait for the emerging powers to accept a greater role (Leal-Arcas 2011).

The key problem identified by most of these scholars is structural: the asymmetric distribution of power, economic interests, and vulnerability among the parties to climate change negotiations favors inaction because those who have the means to address the problem perceive the costs of climate change mitigation to outweigh its benefits (Victor 2011). This argument fits squarely into the rational-choice framework at the heart of the two dominant schools of thinking in international relations—neorealism and neoliberal institutionalism—which share the basic assumption that persons and states alike are rational decision makers; they consider potential costs and benefits when assessing different policy options, including the option of doing nothing. Rational-choice theory predicts that states will pursue the policy option that maximizes their national utility, usually measured in terms of power or wealth. Currently, it seems, the cost-benefit calculations of major players in the global climate governance regime are "net negative"—states believe that the costs of taking action outweigh its benefits.

The strategic nature of this rational approach has led to a wide application of game theory—the attempt to analyze international affairs through the lens of highly simplified game analogies of real-world situations. One of the most interesting recent applications of game theory to global climate change politics suggests that international cooperation depends on, and could therefore be induced by, the reduction of scientific uncertainty concerning the location of a climate tipping point (Barrett and Dannenberg 2012, 2014). A climate tipping point "occurs when a small change in forcing triggers a strongly nonlinear response in the internal dynamics of part of the climate system, qualitatively changing its future state" (Lenton 2011b, 201). The reduction of uncertainty concerning such an occurrence would change the character of the game (or negotiations) from that of a prisoner's dilemma inviting free riding to one of desirable cooperation. Unfortunately, the needed reduction of scientific uncertainty does not seem achievable in the foreseeable future (Lenton 2014).

Some scholars have expressed doubts about the utility of rational-choice models of burden sharing with regard to major global problems such as climate change. Daniel Bodansky points out that the collective-action rationale—everybody is interested in everyone else making binding commitments—does not seem to apply in the case of climate change. The group of BASIC countries—Brazil, South Africa, India, and China—should have a strong interest in having the developed countries adopt binding numerical emission reduction targets, but they strongly resisted the inclusion of any binding numbers in an international agreement debated at COP 15 in Copenhagen (Bodansky 2011). At the same time, the nations of the European Union have been pushing ahead with costly greenhouse gas reductions, knowing that few other nations were doing the same. If these actors are not making rational-choice decisions, what are the thought processes underlying their positions?

In seeking to answer this question, the social constructivist school of thought assumes a very different decision-making reality based on ideas—actor identities, norms, and conceptions of justice and injustice (Albin 2001) in any given context—rather than material costs and benefits. In the context of international climate governance, social constructivists tend to focus on various processes of social meaning making (Miller and Edwards 2001; Miller 2001; Pettenger 2007; Liverman 2009) rather than on explanations of cooperative or noncooperative negotiation outcomes. Among the exceptions are Timmons Roberts and Bradley Parks (2006), who offer a detailed account of the role of opposing perceptions of climate justice as barriers to a multilateral agreement, and Robyn Eckersley (Eckersley in Reus-Smit 2004, chap. 4), who attributes the failure of international climate change treaty making to differences in regulatory ideals, moral norms, and identities.

Related to the social constructivist work on justice and equity in the climate governance regime is a ballooning literature on climate change ethics and climate justice (Ringius, Torvanger, and Underdal 2002; Okereke 2010; Gardiner 2011; Pickering, Vanderheiden, and Miller 2012; Page 2013; Grasso and Markowitz 2015). Grounded in political and moral philosophy, this scholarship explores ethical norms that could guide the political distribution of burdens (e.g., mitigation and finance obligations) and of rights and benefits (e.g., financial support for adaptation, technology transfer) among parties to the global climate regime. To a lesser extent, it also explores more general moral questions the regime has to grapple with, including an appropriate goal for climate governance or how to avoid or compensate harm (loss and damage) caused by climate change impacts.

Recently, several additional topics have enriched the research on climate change politics. These include the interaction between domestic and global climate change politics (Harrison and Sundstrom 2010; Hochstetler and Viola 2012) and the question of what alternative features of the international regime might make cooperation more likely (Bernauer 2013). Further, the emerging powers have become more central to climate change politics and scholarship (Hurrell and Sengupta 2012; Hallding et al. 2013). Research about their role in the negotiations often has neorealist undertones, emphasizing national interests, great power competition, and geopolitics (Brenton 2013), but some scholars have also investigated the identities and preferred norms of the emerging powers in a climate change context (Hochstetler and Milkoreit 2014). Though it might not have been obvious from an outsider's perspective, the emerging powers, especially China, were key to a successful outcome in Paris. In the lead-up to COP 21, there was a flurry of bilateral diplomacy and collaboration on climate change not only between the United States and China, but also between the United States and India. For many negotiators, the emergence of a Group of 2 (the United States and China), with a range of serious domestic climate change policy initiatives by both nations, was the most important source of optimism that an agreement was possible in Paris. As long as the two largest emitters of greenhouse gases were on the same page, real progress was imaginable—a massive shift for the better since the Copenhagen disaster. This optimism benefited the negotiation process, pressuring more reluctant players like India to get on board (or, rather, get out of the way). Never underestimate the role of psychology in international diplomacy.

To summarize, the vast international relations literature paints an increasingly complicated and inconclusive picture of the state of international climate change politics. The Paris Agreement has temporarily silenced pessimism concerning the utility of the UNFCCC process. But that does not mean that any of the challenges outlined above will disappear. Collaboration is possible, but only at certain times and under certain conditions. Actors seem rational, at least sometimes, but they also respond to normative pressures. The global climate governance regime is becoming more complex, but not necessarily more effective with regard to slowing climate change. Politics require patience but nature moves on.

All of these strands of research within international relations rely on important, implicit assumptions about the thought processes of decision makers that lead to observed behavior. Rational-choice scholars assume that decision makers perform a mental calculation of costs and benefits; they attribute causality to the system structure that presumably presents action opportunities with calculable costs and benefits. Social constructivist scholars, for their part, assume that decision makers are motivated by a prescriptive norm that applies to their identity and situation; they attribute causality to ideas. Neither camp has so far seriously considered that it is the mind of a decision maker that assesses the system structure and produces ideas.

By "rational-actor assumption," I mean the assumption that all political decisions are ultimately rational, operationalized through a utility-maximizing calculation of costs and benefits. Taking this assumption seriously raises a whole host of interesting and challenging questions that could be answered empirically. Who, for example, is involved in conducting cost-benefit analyses (CBAs) for a state delegation in the global climate change negotiations? Given major scientific uncertainties concerning future climate impacts on economies and societies around the world, what are the source data for CBAs? And what types of costs and benefits are taken into consideration, in particular when it comes to nonmonetary values such as human life, biodiversity, the existence of an island or a cultural site?

To my knowledge, no empirical effort has been made to answer these questions. Further, how can one explain the increasing inconsistencies between existing cost-benefit analyses of action on climate change, on the one hand, and the persistent negotiation positions of major actors, on the other? A range of respectable economic analyses indicate that action on climate change would have relatively minor economic costs compared to significant benefits (Stern 2007; Hanemann 2008; Rogelj et al. 2013). These well-known, although not undisputed, analyses favor action on climate change, but they have not yet influenced the negotiation positions of major negotiating parties. Why not?

Addressing any of these issues would take researchers into the domain of cognition—the mental processes that underlie the beliefs, decision processes, and negotiation behavior of individuals and groups participating in global climate change politics. What do these people believe? How do they reason? What rules govern their thought patterns?

To sum up, the theories of very different schools assume that individuals and groups have similar or stable cognitive patterns across different situations. The theories might differ regarding the cognitive elements and processes they consider to be important drivers for human behavior, but all take the existence of cognitive processes for granted. It is not unreasonable to expect that each theory intuitively captures one particular type of cognitive process and that multiple cognitive processes can coexist.

Taking the assumption of stable cognitive patterns seriously, the chapters that follow combine a set of old and new empirical tools to explore the cognitive elements and processes that take place in the minds of decision makers when faced with the challenge of developing a cooperative global regime for climate change. They present and examine evidence that supports both rational-choice and social constructivist approaches, and even suggest that cost assessments, identity definitions, and normative ideas are inextricably linked to one another.

Global Environmental Governance

A research field with a topical focus on global environmental change, global environmental governance (GEG) comprises conceptual approaches to governance and politics that are broader in scope than those of international relations scholarship.

One of the most productive features of GEG research is its embrace of a range of political actors that are largely ignored by international relations scholars, with their natural focus on the interactions between national governments. These actors include nonstate entities (e.g., nongovernmental organizations, corporate and industry groups, individuals) and substate entities (e.g., regional and municipal governments, city networks) that are often organized transnationally (Bulkeley et al. 2014). Exploring the goals, strategies, sources of power, and impacts of these diverse actors offers a broader perspective on global climate governance, one that complements and adds complexity to the insights provided by international relations. This broader perspective reveals multiple modes of governance that are very different from those of formal cooperation of states in multilateral agreements, modes that are often private or informal, voluntary, and geo-graphically or sectorally distinct. Although there are challenges when it comes to measuring the impact or success of these more diverse actors and the institutions they create, GEG literature offers important insights concerning the sources of change at the political micro, meso, and macro scales. For example, some fascinating work that links political economy approaches to global governance illuminates the multiple ways in which markets, market-related institutions, and the norms of actors wielding structural power affect and are affected by political processes (Levy and Newell 2004; Falkner 2008).

With a deep interest in local and regional, in addition to national and global, political processes, global environmental governance scholarship greatly expands the scope of research on specific climate change–related policy issues, such as reducing emissions from deforestation and forest degradation in developing countries (REDD+) or adaptation, and their community-based implementation (Armitage 2005; Sandbrook et al. 2010; Gruber 2011). Concepts like resilience, socio-ecological systems, learning, and participatory processes are common in this research, linking politi-cal concepts to other social and even natural science disciplines, includ-ing sociology, ecology, and geography (Adger, Arnell, and Tompkins 2005; Adger et al. 2009; Wang and Blackmore 2009).

Much of GEG research takes place in the normative context of sustain-able development. The Earth System Governance (ESG) Project, which includes many GEG scholars, acknowledges this normative driver explic-itly, focusing on questions of political, especially democratic, legitimacy, equity, and social justice.

Many global environmental governance scholars have grown interested in the role of knowledge and knowledge production, social learning, val-ues, and beliefs in environmental governance processes (Collins and Ison 2009; Tschakert and Dietrich 2010; Hegger et al. 2012; Zia 2013). What decision makers, stakeholders, and natural resource managers know, how they perceive and construct mental models of the world around them, and how they change their views and values through social learning are increasingly considered essential questions of environmental governance, and therefore important subjects of GEG research. Acknowledging different forms of knowing such as indigenous, cultural, or artistic knowledge sys-tems, GEG scholars are beginning to explore ways of bridging these forms with scientific knowledge (Rathwell, Armitage, and Berkes 2015). All of these trends reflect a turn toward the study of cognitive processes, although without using the terminology, concepts, and tools of cognitive science. Not surprisingly, scholars pursuing this line of inquiry often struggle with developing processes and methods to enable them to investigate mental processes (Blackstock, Kelly, and Horsey 2007; Gidley et al. 2009; Jones et al. 2011, 2014).

Given the roots of global environmental governance and the multiple interactions between social and natural systems at the global scale, it is also not surprising that GEG scholars have recently taken a keen interest in the Anthropocene, the proposed epoch marking the global and lasting impacts of human activities on Earth systems. The emerging field of Anthropocene governance scholarship contemplates whether this new concept is fundamentally changing the rules of the political game, requiring new forms of thinking, or new kinds of governance mechanisms and institutions (Biermann 2014; Pattberg and Widerberg 2015). However, so far, truly novel proposals for global environmental governance have remained elusive with the exception of John Dryzek's argument (2014) for "ecosystemic reflexivity" as a guide for rethinking path-dependent institutions in the Anthropocene. What that means for the future practice of global environmental governance, including the UNFCCC, remains to be seen.

Science and Politics

The relationship between science and politics, on the one hand, and between science and policy, on the other, raises a number of fundamental questions about the nature of society, knowledge, and decision making, questions that have been the subject of vast bodies of theorizing over the last century. Beyond the most general question of the role of science in society, or the nature of the social contract, two domains of research have emerged: policy for science (i.e., the political institutions set up to govern scientific activity) and science for policy (i.e., the interactions between science and policy that supposedly inform and guide political decision making). Here I am most interested in the latter, which includes challenging questions about the appropriate role and behavior of scientists in policy processes and the design of scientific institutions informing global political processes, such as the Intergovernmental Panel on Climate Change (IPCC) and the UNFCCC. An institutional innovation when it was established in 1990, the IPCC has become the gold standard for new global governance regimes that require scientific input, such as the Intergovernmental Platform on Biodiversity and Ecosystem Services (IPBES). Over the last decade, extensive scholarship on these questions has made huge strides, moving from what I call the "linear model" to an exploration of the complex interactions between scientists and policy makers.

For most of the twentieth century, science and politics were defined as two separate spheres, adhering to different rules and committed to different values. These separate worlds were connected by a one-way flow of information from the sphere of science (truth seeking) to the sphere of politics (value contention), where rational decisions were expected to be made based on scientific facts. Mike Hulme (2009, 102–104) has called this linear relationship the "technocratic model." Based on this conception of the two spheres as separate, the theory of boundary organizations explores how specific organizations can create interactions, practices, and so-called boundary objects that can help bridge this divide between the spheres and facilitate knowledge transfer (Guston 2001).

To a large extent, the relationship of the IPCC and the UNFCCC can be characterized as linear. The IPCC aggregates and synthesizes all available climate science in major reports that are presented to the decision-making community in the UNFCCC. The IPCC "Assessment Reports" are supposed to be policy relevant but not policy prescriptive, that is, to inform necessary decisions but not to advocate for certain policies. One key feature of the IPCC process, the "Summary for Policy Makers," a condensed version of each Assessment Report's key insights, does not fit the linear model, however, requiring *joint* approval by policy makers and scientists. This interesting governance twist often results in an epic battle over words and phrases, clear evidence of a significant shortcoming of the neat, but unrealistic linear model of the science-policy interface (Hulme and Mahony 2010).

Over the last two decades, a growing body of scholarship has recognized that science is a social activity, conducted by human beings with values, incentives, and emotions, and that there are multiple complex interactions between the spheres of science and politics. In recognizing as well that no single, comprehensive model can describe this complex reality, these scholars have created conceptual tools to better study and understand it, most notably, "wicked" problem analysis (Rittel and Webber 1973), post-normal science (Funtowicz and Ravetz 1993), participatory science (Blackstock, Kelly, and Horsey 2007), knowledge coproduction (Jasanoff 2004), and knowledge brokering (Meyer 2010).

This messier perspective offers a range of explanations for the challenges experienced by the IPCC and its contributing scientists over the last few years. Instead of focusing on matters of institutional design at the science-policy interface, scholars have explored the professional and ethical obligations of scientists (Gamson 1999; Lackey 2007). Should scientists be allowed to express support for certain policy solutions, in other words, should they become advocates under certain circumstances, or should they remain neutral providers of knowledge, no matter what?

Related to the questions about the science-policy interface are issues of science communication, public belief formation, and the politics of meaning making. In the climate change arena, politically motivated skepticism (Smith and Leiserowitz 2012; Matthews 2015; Torcello 2016) and the strategic organization of climate change denial (Jacques, Dunlap, and Freeman 2008) have received a great deal of attention. Beyond the immediate effect of these phenomena on domestic climate change politics and policy making in the United States and their indirect effects, through congressional climate change politics, on international diplomacy, they have worked to weaken the social contract between science, government, and the public, undermining public trust in science and scientific knowledge. When scientific knowledge no longer has a special status in policy debates that is distinct from value-based opinions, the nature of policy making and society itself changes significantly.

Although necessary to create a deeper understanding of the dynamics between the spheres of science and politics, this movement toward questions of advocacy, beliefs, politics, and power is still too narrow and inward focused. Scholars have addressed almost exclusively the role of scientists and their obligations to act "responsibly" if they wish to have an impact in the world of politics. There are widely diverging views on what it means to be responsible, ranging from the moral obligation to speak out on policy issues to the need to remain objective, neutral, and out of policy debates, but the arguments all revolve around an appropriate code of conduct for scientists, who investigate issues with significant and sometimes frightening human implications. This scholarship and much of the media coverage seem to suggest that only the scientists can fix the communication mess by figuring out how to interact with the political sphere—how to design boundary organizations, how to offer scientific advice, whether or not to advocate, and how to conduct science in a world full of urgent problems. These arguments implicitly assume that the "right" kind of behavior of scientists will lead to rational, "technocratic," or, more generally "desirable" responses on the policy-making side. Thus we find ourselves back where we started—at the rational-actor assumption.

Much of *Mindmade Politics* will argue that a narrow focus on the role of scientists unduly constrains the questions we ask, the data we seek, and consequently the explanations we offer for a poorly functioning science-policy interface. And it will suggest that, to arrive at a more productive relationship and the needed understanding to explain and support such a relationship, scholars must shift their analytic lens away from the providers of scientific knowledge to the recipients of this knowledge—political actors.

A question that has not yet been asked is what part of the available scientific knowledge do political actors actually understand, absorb, and actively use when determining and pursuing their political interests? What is the link between scientific knowledge and political or national interest? In the context of climate change, how much climate science do UNFCCC negotiators understand and actively use when formulating and advancing their negotiation positions? Does the science inform cost-benefit analyses? Does it support certain norms or moral assessments? Asking what part of a scientific message has penetrated the minds of decision makers, rather than how it was produced and whether scientists advocated for it, points scholars inevitably toward the domain of cognitive science.

Summary

Conventional international relations scholarship has hit a wall in its analysis of the politics of global climate change and would benefit from a serious investigation of the nature and origins of beliefs about costs and benefits, identities and norms, what they actually are, and how they change. Global environmental governance scholarship raises the profile of cognitive variables such as knowledge, beliefs, and values but often does so without the empirical tools to investigate these phenomena rigorously. In researching the transfer, spread, and use of knowledge in the relationship between science and politics, scholars have yet to look at the cognitive realities of these communication processes.

But the rich literature on international politics relies on important assumptions concerning the mental patterns that influence decision making. I treat these mental patterns as central variables in my analysis—cognitive elements and processes that shape how political actors think, feel, and make decisions about global climate governance.

Collectively, the three bodies of research I have summarized in this chapter point to the following set of basic cognitive entities as determinants in global climate change politics:

• Mental representations of costs and benefits related to different policy options in a given situation, which enable the calculation of the expected net utility of a particular political decision or behavior (this reflects the rational-choice assumption); in other words, the mental representation of structural opportunities for and constraints on political agency,
• Mental representations related to different actor identities (in- and out-groups),
• Mental representations related to norms of justice and possibly other norms,
• Mental representations concerning the available scientific information about climate change, including mental models of environmental and social change.

Established theories of cognitive science suggest that the different cognitive elements do not exist independently but tend to be grouped together in coherent structures to form images, narratives, or discourses. Although these structures provide the foundation for individual and collective reasoning and decision making, it is unclear whether one can make general statements about them and the cognitive dynamics they produce across different issues and situations. There is as yet no political theory based on cognitive, rather than material or ideational, structures.

That said, I initiated this research project with the assumption that each of the major theories mentioned above has equal validity—that all of the cognitive assumptions they make are relevant for political decisions regarding international cooperation on climate change. Indeed, the various mental processes they each focus on—cost-benefit analyses, identity conformation, or normative reasoning—may all take place at the same time (parallel processing). Depending on the context and the decision maker, some elements might be more important than others. This opens up the possibility that different theories of international relations are linked rather than opposed to one another, which raises an interesting question. Is there a mechanism that facilitates decision making based on rational-choice, identity, and normative considerations all at the same time? *Mindmade Politics* will answer this question with a resounding yes.

3   The Promise of a Cognitive Approach

Over the last decade, scholarly interest in cognition and even more so in emotion has soared across the social sciences. Scholars of international politics, however, have been slow to join this trend, and when they have, their use of cognitive theories and methods has been limited. In this chapter, I introduce the basic cognitive theories and concepts that inform the research for *Mindmade Politics* and touch on the questions these theories raise for studying climate change politics ("What Is Cognition?"). With a summary of existing cognition-oriented research in political science and international relations, I then show how far these disciplines have come in adopting and utilizing insights about the human mind, highlighting how much more potential there might be for innovative research strategies at the intersection of these disciplines ("Cognition and International Politics").

Many other social science disciplines, in particular, moral psychology as well as the communication and decision sciences, have developed research programs exploring mental processes in response to climate change. I review this quickly growing field of scholarship and its key findings ("Cognition and Climate Change"), paying particular attention to cognitive dynamics when dealing with some of the special characteristics of climate change ("Climate Change: A Unique Cognitive Challenge?").

In closing, I point out and discuss the limitations of a cognitive approach to social scientific research ("Limits of a Cognitive Approach") before presenting a framework for a cognitive analysis of global climate change politics in the final section ("A Cognitive Framework for Political Analysis").

What Is Cognition?

Cognition refers to all processes of thought and feeling, including phenomena such as knowing and acquiring knowledge, awareness, perception, reasoning, memory, judgment, and decision making. These processes consist of mechanisms that interact across molecular, neural, psychological, and social levels (Thagard 2010b). Hence cognitive science is the interdisciplinary study of mind and intelligence, drawing on psychology, artificial intelligence, neuroscience, linguistics, and philosophy.

Many disciplines study cognition, but they define and use the term differently. Dominant cognitive theories take an information-processing view of cognition, comparing the brain to a computer. In close alignment with this information-processing approach, I define cognition as the elements, structures, and processes of individual and collective thought and feeling. Cognitive analysis is thus concerned with what is going on in a person's head and the intersubjectively shared ideas, meaning systems, and emotions in a group.

Theories of Mental Representation

Theoretical neuroscience has taken two major approaches to individual cognition. Earlier theories take a verbal processing approach, conceptualizing cognition as based on the application if-then rules (Thagard 2005, chap. 3); only one rule operates at a time, and series of rules result in decisions. In contrast, later theories—connectionism—take a neural network approach, conceptualizing cognition as processes of a complex network, in which individual cognitive elements are nodes that can be activated by links between them. Because knowledge is coded in neural network structures by simultaneous activation of several nodes, learning and cognitive change require a structural reconfiguration of the network (Antal and Hukkinen 2010, 938); information spreads in nonlinear ways through parallel processing (Thagard 2005, chap. 7), and the central process for making decisions or solving problems is emotional coherence (Thagard 2000). The distinction between these two approaches has important methodological implications, for example, when developing computational models of decision making. Recently, however, the verbal processing and connectionist approaches have been converging around the understanding that the mind uses both rule-based and neural network operations (Eliasmith and Anderson 2004).

Building Blocks of Cognitive Theory

The basic ontological entities of cognition are mental representations, structures, and processes. There are several basic kinds of mental representations, such as concepts, beliefs, goals or motivations, images and representations of events. Concepts usually correspond to single words or terms that stand for something in the world, for example, chair, parents, or climate change. They are relational in that they are linked to entities in the material world and to numerous other concepts; a concept only makes sense—is meaningful—in the context of those relationships. For example, the concept of parents is not only linked to two specific people in one's life—one's mother and father—as well as other parents one knows, but also to large set of other concepts, such as child, relationship, responsibility, care, family, generation, and, to a lesser and lesser extent, marriage.

Beliefs are propositions or convictions; they are statements about the world, such as "Climate change is a hoax" or "Fast food is bad for your health." Causal beliefs—the most important type of beliefs—provide the foundation of goal-oriented human behavior. Goals are desired states of the world that orient and drive human behavior. In order to achieve a certain goal, an individual needs to be able to identify the actions and processes that can lead to the desired outcome.

Mental structures refer to the linkages and relationships between mental representations. From a connectionist point of view, these structures are best conceptualized as networks in which the nodes are individual mental representations with various connections (links) between them. Clusters of concepts form beliefs, images, or other cognitive structures. Each of these structural elements is part of a larger cognitive network and becomes activated in particular contexts.

Mental processes include making decisions, solving algebra problems, daydreaming, and assessing risks. A particularly interesting set of mental processes for the study of politics includes the acquisition, change, and abandonment of mental representations and mental structures, for example, adding new concepts, changing beliefs or worldviews, and replacing existing concepts.

While this typology of cognitive elements answers the question, What kinds of things exist in the mind?, it barely begins to account for the semantic content of cognition. Where does meaning come from? According to connectionist theories, meaning emerges both from the connections between multiple cognitive elements and from their relationship to entities in the material-social world (Markus and Hamedani 2010). Thus my mental representation of climate change consists of a large number of concepts (e.g., the greenhouse effect, temperature change, Arctic summer sea-ice melt, food scarcity, climate tipping points, international negotiations, Kyoto Protocol, solar power), images (e.g., glacier retreat in a time series of photographs), goals (e.g., teaching students) and processes (e.g., imagining how future climate change might affect my hometown). Each element is linked to many others, forming a specific network structure, and each relates to physical realities and events (e.g., the experience of unusual weather, or images of an extreme snowstorm in Boston combined with news reports linking the event to climate change), information regarding scientific findings in academic journals, or conversations with colleagues about global climate governance. When making a decision (e.g., what kind of course I am going to teach next term), all these elements and their linkages play a role.

This systems ontology of cognition, though based on the biological processes and structure of the brain and the human nervous system, needs to be distinguished from them. Neuroscientific descriptions of cognition involve molecules, brain cells (neurons), synapses, neurotransmitters, neural circuits, and firing, as well as neuroimaging. But the growing body of theoretical cognitive science serves as a bridge between neuroscience as an interdisciplinary subfield of biology and the social sciences, which seek to leverage the advances in our collective understanding of the brain for studying social phenomena such as belief formation, decision making, and behavior.

Cognition and Emotion

Cognition and emotion are inextricably linked—feeling is integral to knowing. Building on advances in the cognitive sciences over the last two decades (Damasio 1995; Loewenstein et al. 2001; Vohs, Baumeister, and Loewenstein 2007; Duncan and Barrett 2007; Scholl 2013), some theorists argue that previous views of cognition as computational processes of deliberative coherence are incomplete. Putting forward a theory of cognition as a process of "emotional coherence," Paul Thagard (2006, chap. 2; 2008) suggests that emotions are mental states that cannot be separated from the cognitive elements described above. Concepts can have emotional valences (e.g., positive emotions related to the idea of a long vacation, or negative emotions related to the concept of death), as can beliefs (e.g., "Climate change is real" can evoke fear and guilt) and goals (e.g., thinking about getting your driver's license or a promotion makes you happy). Emotions are also involved in cognitive processes, such as rejecting or revising beliefs. Accepting a new belief based on coherence feels good; if a new belief is incoherent with our existing beliefs, we find it irritating. Thagard assumes that decisions are based on satisfaction of multiple constraints, taking into consideration both the cognitive acceptability of a mental representation and its emotional valence. Based on this view, emotions are intrinsic components of all cognitive elements, structures, and processes; any cognitive theory needs to account for the emotional content of cognition.

This integrative or hybrid view of cognition as inseparable processes of thought and feeling is, however, still contested in many disciplines, including the cognitive sciences. The prevalent view of cognition and emotion defines them as separate but interacting systems, each with functionally specialized areas in the brain. Although this view has already evolved from Robert Zajonc's position (1980) that emotion is primary and independent of cognition, and from Richard Lazarus's position (1982) that emotion is secondary and dependent on cognition, recent work increasingly emphasizes that "there are no truly separate systems for emotion and cognition because complex emotional-cognitive behavior emerges from the rich, dynamic interactions between the brain networks.... [T]he neural basis of cognition and emotion should be viewed as strongly non-modular" (Pessoa 2008, 148).

Individual and Collective Cognition

Cognition takes place in the brains of individuals, but all social behavior depends on the ability of social groups to attribute shared meaning to objects, behaviors, events, and words (language).[1] Fundamental to the

1. Parts of this section have been published previously in Milkoreit and Mock 2014.

existence and development of human societies, collective meaning making is at the heart of political decision making. But can we actually attribute beliefs to social groups—is there such a thing as collective cognition? If not, because there is no group brain, how can we conceptualize the processes and results of collective meaning making? Closely linked to the question of collective thought is that of collective emotion. Roland Bleiker and Emma Hutchison (2014, 491) have called "theoriz[ing] the processes that render individual emotions collective and thus political" the central challenge of emotion-related research in international relations.

As mentioned earlier, cognition is a multilevel process, comprising interactions between molecular, neural, psychological, and social mechanisms (Thagard 2010b). The idea of multilevel interacting mechanisms is useful to explain the cognitive relationship between individuals and groups. Individual-level mechanisms (i.e., molecular, neural, and psychological processes) interact with group-level mechanisms (i.e., communication and sensory interaction) to create the bonds that constitute and hold a group together.

The key to collective cognition are the individuals who make up their respective groups and who think of themselves as members of those groups (Thagard 2010b, 274). Individual group members have many kinds of mental representations that contribute to their understanding of the group. At least three kinds can be distinguished: identity-related concepts about the self (e.g., individual values that motivate group membership), concepts about other group members (e.g., their attributes as members), and concepts about the group as an entity (e.g., the purpose and activities of the group). Individuals acquire and change group-related beliefs, especially those concerning the group as an entity, through interactions with other people (both members and nonmembers) and through group-related experiences. Examples of such experiences include the use of shared resources or property, for instance, certain spaces and office buildings, or the collective organization of or participation in events. This process of social communication and physical and sensory interaction facilitates the convergence of mental representations related to the group in the minds of group members and nonmembers alike. The process works both ways: while receiving information about and developing an understanding of the group, an individual also contributes to other individuals' mental representations and experiences of the group. The nature of the group depends on this recursive process between individual cognition and social interaction. Bridging multiple levels, this process does not create a shared or single mind, but the existence of very similar cognitive elements and structures in the multiple minds of group members.

Social psychologist Naomi Ellemers (2012) and other scholars of social identity theory (e.g., Ellemers and Haslam 2016) support this understanding of the relationship between individual and collective cognition. Writing about intergroup conflict, Ellemers (2012) explores the conditions under which the group self—thinking about oneself as a member of a group—can become more important than the individual self.

Understanding collective beliefs therefore requires, first of all, understanding how individuals envision themselves as group members and the emotions, values, and meanings they attach to this membership (Thagard 2010a, 274; see also Tajfel 1982, 2–3). Second, it requires understanding how communication and physical interactions of group members with one another and with the group's social-material environment allow individuals to share ideas and emotions, enabling both individual and collective cognitive change (e.g., acquiring or revising shared beliefs). The social communication processes that lead to shared beliefs operate in both directions. Much of individuals' belief systems is received from the outside, severely limiting the ways they can envision themselves as group members. At the same time, an individual's actions and use of speech can affirm, challenge, and change group-related ideas of other group members, ultimately affirming or changing the group itself.

The argument that individual and collective beliefs about the group mutually influence each other raises the question whether and under what conditions one side of this interaction might be dominant. Do we as individuals have the cognitive freedom and agency to develop and change our own beliefs about our groups (and the world) or is much of what we believe about the groups we are part of imposed on our minds by the groups? Do we have the power to pick and choose certain ideas about our groups, or do we have to accept the whole package? The answer depends to a large extent on the type of the group (e.g., your local hiking club vs. your nation), its developmental stage or age, and your position in it (e.g., whether you are the CEO of the company or an employee in the accounting department). More generally, the network of mental representations that describe and give meaning to a group identity has to be emotionally coherent and fairly similar across many members' minds. Picking and choosing certain parts while rejecting others might not be possible if the rejected pieces include important collective-identity elements. Removing core elements from an emotionally coherent system of ideas could reduce or even destroy this coherence. This network of ideas has developed over the entire history of the group, providing the basic schema that constitutes the group and gives it stability. A new group member usually has to adopt the whole network of ideas as a coherent whole; rejecting parts of that network would imply staying outside of the group. For example, a Swiss national cannot reject the idea of neutrality as part of what it means to be Swiss, but might very well object to the idea of Swiss cheese as a national symbol (Milkoreit and Mock 2014, 174–180).

Conceptualizing the link between the individual mind and the collective beliefs of the group as mutual constitution is reminiscent of Anthony Giddens's structuration theory (1992), which suggests that there is a mutual relationship between social structures and agents. Structures both constrain and enable meaningful behavior; agents can use existing structures to create social change. The structural conditions—the existing network of ideas forming the group identity—at any point in time serve as a stable reality against which ideas for alternative desired conditions can be presented. These ideas can destabilize existing institutions, norms, or identities, but only because these structures have been perceived as stable before.

Linking Cognition and Social-Material Realities

Cognitive processes are influenced not only by our interactions with other people, but also by our interactions with the material world of objects—natural and artificial entities, such as plants, air, buildings, or books—and with events, including storms, meetings, terrorist attacks, or extinctions of species. When talking about climate change, objects and events in the natural world are of particular interest: how do we categorize the natural kinds of these (e.g., climate zones), how can we observe change (e.g., the melting of the Arctic summer sea ice), how do we make sense of extreme weather events? And how do things take on certain meanings for us, sometimes different meanings for us than for other people?

A special problem in this context is the role of scientific information. Can abstract scientific information and concepts have the same cognitive influence as, for example, the personal experience of flooding? And is scientific information simply a reflection of the physical, chemical, or biological reality or a social construct? Does it matter that abstractions like average global temperature and climate sensitivity depend on the ability of the human mind to create concepts and meaning around phenomena we cannot directly observe?

A full theory of cognition will have to take all the processes outlined above into consideration: relational processes of meaning making, linking multiple concepts in the mind with things in the world; interactions between individuals and social groups; and interactions between individuals (or groups) and the physical environment (see figure 3.1). These various processes critically depend on how we communicate and physically interact with one another, how we observe with our senses and their technological extensions (e.g., telescopes, cameras, and satellites), and how we interpret what we perceive. All of these processes contribute to the creation of intersubjective agreement on the meaning of objects, events, and social facts.

As the following sections will show, social scientists have begun to build on basic insights about our minds to study well-known social phenomena with a new set of conceptual lenses. Though often lacking the necessary tools and methods to investigate cognitive and affective processes, they have come to recognize the promise and potential value of a cognitive approach to political science, communication and decision sciences, economics, ethics, and (moral) psychology.

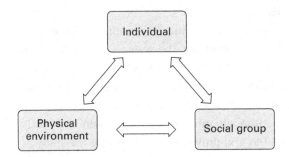

Figure 3.1
Relational cognition.

Cognition and International Politics

What is the relationship between mental processes and political decision making? In his well-known book, *The Political Mind* (2008), George Lakoff argues that the success and failure of conservatives and liberals in the politics of the United States depends largely on their respective abilities to tap into cognitive-affective processes of voters' minds. According to Lakoff's analysis, conservatives are much more successful at this mental engagement, giving them (at least for the moment) a significant advantage when it comes to mobilizing citizens for their cause. Other scholars have used cognitive theory to, for example, deepen arguments for methodological individualism (Turner 2003), stressing the interaction between culture, emotion, and reason (Connolly 2002), or linking emotions to the rational-choice framework (McDermott 2004). There have also been some initial efforts to go beyond the interface of cognition and politics and explore the biology of politics (Hatemi and McDermott 2011), in particular, the biology of conflict (Berns and Atran 2012; Berns et al. 2012). This research seeks to trace the origins of political behavior all the way to the differences in the genetic makeup or brain chemistry and functioning of political actors.

Most of the authors mentioned here adopt two of the major tenets of modern cognitive science. First, they reject mind-body dualism in favor of a brain-based definition of the mind (Gunnell 2007, 705). Second, building on Damasio's groundbreaking work (Damasio 1995, 2003), they acknowledge that reason and emotion cannot be separated but interact as parts of an integrated whole.

International relations as a discipline is only beginning to discover the utility of a cognitive approach to studying the interactions of states or the dynamics of world politics. As I pointed out in chapter 2, traditional analyses of international politics have relied on behavioralism and the formal modeling of decisions based on a set of unrealistic assumptions—the rational-choice framework. Although this approach has been eminently successful in economics and has also proven to be fruitful for the study of certain problems in international relations, it has come up against important explanatory and predictive limits.

Social constructivist theories have sought to counter and complement rationalist approaches by arguing that ideas matter beyond the calculation of costs and benefits. They have focused on certain types of ideas that are often ignored in rationalist work: identities and norms. Actors' self-conceptions and the norms they follow can heavily influence the goals they pursue and the choices they make, sometimes trumping concerns about costs and benefits. A cognitive perspective is very much in alignment with social constructivism—it is about ideas and their influence in international politics. But a cognitive approach asks new questions about the origin of ideas, about the way in which multiple ideas hang together and form systems of beliefs, about how these belief systems influence decision making and finally, about the ways in which belief systems can change over time.

Political cognition is therefore best understood as a set of observable actor-level processes that bridge the material-social world and its given ideational structure (cognitive input), on the one hand, and political decisions and behavior (cognitive output), on the other. Cognition can be conceptualized as the gate through which information about and perceptions of the material-social environment passes in order to lead to a decision or behavior.

Cognition has been the subject of two different research programs within international relations. The first, in the field of political psychology, is the study of political behavior from a psychological perspective. Drawing heavily on cognitive psychology to study the decision rules or "shortcuts" of our minds when dealing with complex problems in various circumstances, political psychologists have found that political actors tend to rely on heuristics, such as availability and anchoring, including affect-driven heuristics (Finucane et al. 2000). They are also subject to framing effects, such as loss aversion when making risk assessments, as Daniel Kahneman and Amos Tversky (1979) famously proved, and other cognitive biases. These findings often supplement those of rational-choice theorists and their assumption that human beings are expected-utility maximizers and explain the frequent deviations of political actors from expected rational decision making (Gross Stein and Welch 1997, 53).

Focusing on the issue-specific content of beliefs, I build on a second research program: the study of the beliefs and (lately) emotions of individual decision makers. Researchers in this program make use of four distinct theoretical tools: (1) the operational code hypothesis; (2) cognitive or conceptual maps; (3) discourse analysis; and (4) image theory.

Developed by Robert Merton (1940), and reformulated with more conceptual rigor by Alexander George (1969), the "operational code" hypothesis suggests that political leaders have specific beliefs about the principles of political life. George developed ten questions that can capture these beliefs; he argued that a person's responses to these questions form a belief system, in which individual beliefs are "bound together by some form of constraint or functional interdependence" (George in Young and Schafer 1998, 70). Operational codes are issue specific and subject to change over time (Renshon 2008).

David Axelrod's publication of *Structure of Decision* in 1976 triggered a wave of research on cognitive or conceptual maps. Scholars developed sophisticated approaches to mapping the beliefs of individual policy makers and decision-making bodies. Although most studies were based on text analysis, some verified their mental models through simulations with policy makers (Bonham et al. 1988). The core strength of these early maps is their reflection of causal beliefs, assumed to be the basis of decision making (Astorino-Courtois 1995). Matthew Bonham (1993) sought to connect cognitive mapping and international negotiations, arguing that the tool can reveal the assumptions of negotiators, map the effect of individual proposals, and identify common ground.

Michael Shapiro, Matthew Bonham, and Daniel Heradstveit (1988) had proposed an alternative approach to cognition, arguing that "discursive practices" are historically determined constraints for individual cognitive processes. Their work speaks to whether the individual mind matters at all when its contents are determined or at least influenced by the social context. A discursive practice (frame) is conceived as external to the individual mind and treated as the agent in the causal model. More generally, "discourse analysis" has become a prominent constructivist tool in the study of global environmental politics (Hajer 1996). A discourse is "a specific ensemble of ideas, concepts and categorizations that are produced, reproduced and transformed in a particular set of practices and through which meaning is given to physical and social realities" (Hajer 1996, 44). Discourse analysis seeks to understand the dynamics that can lead to the dominance of one discourse over others or to the change of social institutions. The formation of discourse coalitions plays a major role in this process (Maguire and Hardy 2009; Maguire 2004), as does the strategic use of narratives (framing) and the availability or control of communication resources or agendas (structural power). Major weaknesses of discourse analysis include its inability to explain the origins or success of discursive content and the lack of generalizability across cases. Further, there remain important questions concerning the nature and causal power of discourses and ideas more generally.

Another body of research studies images as relatively stable mental structures that influence decision making (Cottam 1986; Herrmann et al. 1997). This work is based on the notion of schemata as organized clusters of concepts. Richard Herrmann and colleagues (1997) connect image theory to gestalt theory, arguing that an image forms an integrated whole rather than a collection of separate and independent parts. International relations scholars focus on images of other countries in the minds of individual decision makers based on three sources: goal compatibility or incompatibility, power differences (i.e., the potential for agency), and cultural status (identity). Michele Alexander, Shana Levin, and P. J. Henry (2005) link image theory to social identity and social dominance, based on ideas very similar to those of cultural theory (Douglas and Wildavsky 1982).

Since the turn of the century, researchers in political science and international relations have also taken a growing interest in emotions. Given their inseparability from cognitive processes, emotions are tied up with all cognitive variables political scientists find relevant to political behavior, in particular with concepts of identity and justice (Mercer 2014). Some of the most insightful works include Drew Westen's exploration of subconscious emotions in American politics in *The Political Brain: The Role of Emotion in Deciding the Fate of the Nation* (2008), George Marcus's analysis (2000) of emotions as tools for evaluating political circumstances, and a number of edited volumes such as *Affect Effect* (Marcus et al. 2007), *Political Emotions* (Staiger, Cvetkovich, and Reynolds 2010), and *Emotions in Politics* (Demertzis 2013). There is also a growing literature that considers cognition and emotion—whether defined as variables or processes—important for understanding, explaining, and possibly even predicting political behavior in the international system. For example, Neta Crawford (2000), Rose McDermott (2004), and Jonathan Mercer were among the first to introduce the idea of affective rationality into the field of international relations. Mercer (2005; 2010) argues that emotional cognition is an "assimilation mechanism" for new information into existing belief structures and plays a major role in risk assessments. Hutchison and Bleiker (2014, 496) describe the role of emotions in politics as "forms of insight and sources for political decision"; they even argue that emotions "have now become one of the most exciting theoretical and empirical research areas in international relations" (495).

A small set of questions emerges at the heart of this new field of research, which is likely going to be the subject of exciting debates in the coming years. The most important question concerns the link between individual and collective emotions—what makes emotion a collective phenomenon with effects on the processes and outcomes of world politics (Hutchison and Bleiker 2014; Mercer 2014; Crawford 2014; Linklater 2014; Reus-Smit 2014)? Is it even appropriate to speak about collective emotions, or should researchers instead turn their attention to the individual, bodily, somatic experience of emotion (McDermott 2014)? To these questions, I would add the following three. What is the relationship between cognition, emotion, and rational choice? How does cognition relate to structural realities and the concept of power? And can cognitive approaches offer new empirical opportunities for social constructivist work?

These questions will not be easy to answer. Some of the key challenges of this emerging field of research are methodological. There are few reliable research tools for identifying the beliefs and emotions of political actors and for tracking these over time; this paucity of tools presents significant obstacles to high-quality empirical work that could in turn advance theory. More generally, what is needed is a stronger conceptual bridge between cognitive science and theoretical accounts of beliefs, emotions, and decision making to create consistency across multiple disciplines from neuroscience (e.g., the role of neurotransmitters such as oxytocin) to cognitive science (e.g., computational modeling of network-based information processing in the brain), psychology, and political science.

Cognition and Climate Change

In a spectacular explosion of scholarship on the mental and emotional processes involved in climate change–related belief formation, decision making, and behavior, communication scholars, decision scientists, and psychologists have, over the past two decades, identified and compared beliefs about and attitudes toward climate change among different segments of the public, often drawing out lessons for communicating about and generating public support for policy on climate change. One of the best-known examples of this work is the ongoing research of the Yale Project on Climate Change Communication (Leiserowitz 2006; Smith and Leiserowitz 2012; Roser-Renouf et al. 2015). Some of the communication challenges this research has identified include people's lack of experience of climate change, its invisibility, distant impacts, complexity, and uncertainty, and the general perceptual limits of people in dealing with climate change (Moser 2010).

Comparative analyses have established both differences and similarities in climate change–related beliefs and attitudes across cultural, social, and geographical groups (Lorenzoni et al. 2006; Wolf and Moser 2011; Brechin and Bhandari 2011; Kvaløy, Finseraas, and Listhaug 2012; Smith and Leiserowitz 2012). Although there are often major differences in beliefs and concerns within societies, Beatrice Crona and colleagues (2013, 520) have found a surprising pattern of similarities across very distinct cultural groups around the world, which they call a "global, cross-cultural mental model around climate change." According to this mental model climate change is manmade, creates extensive changes in the natural environment, is associated with more extreme weather events, and can have negative health effects on humans.

More recently, scholars have become interested in the relationship between the sensory experience of a climate change–induced phenomenon, such as flooding, and climate change–related beliefs (Dessai et al. 2004). They have observed influences in both directions (Myers et al. 2013). Thus, if you believe in climate change, you interpret your experience in ways that confirm or strengthen your beliefs (Zaval et al. 2014; Spence et al. 2011), but if, on the other hand, your experience contradicts your past climate change beliefs in irreconcilable ways, you are likely to change those beliefs (Reser, Bradley, and Ellul 2014).

Beyond the quantitative analysis of knowledge, beliefs, and public concern about climate change, a second body of research explores people's cognitive and behavioral responses to the problem (Norgaard 2009). Numerous research programs describe cognitive-affective coping mechanisms people use to deal with or protect themselves from difficult or threatening information contained in climate change science. These mechanisms include nonengagement and distancing, denial and skepticism, or engagement in social movements (Roser-Renouf et al. 2014). Kari Marie Norgaard (2011) has developed a sociological account of the social mechanisms complementing and embedding many of these psychological factors in community life. Her theory of socially organized denial suggests that people make an effort to keep disturbing information at a distance in order to (1) avoid negative emotions, such as fear, guilt, and helplessness; (2) follow cultural norms, such as not raising difficult subjects in conversation that could embarrass or even humiliate interlocutors; or (3) maintain positive conceptions of individual and national identity. Societies develop a repertoire of such techniques to ignore disturbing problems and create narratives in which "everything is fine." In short, Norgaard's theory suggests that, rather than lacking information, people do not want to know about climate change for a range of psychological and social reasons (Norgaard 2011; Norgaard in Dryzek, R. Norgaard, and Schlosberg 2011, chap. 27).

Finally, building on past work in social and environmental psychology, scholars explore multiple factors shaping or discouraging proenvironmental behavior (e.g., climate mitigation) and behavioral change generally (Blake 1999; Kollmuss and Agyeman 2002; Gifford 2011).

Diving deeper into some of the intra- and interpersonal factors that shape decision making on climate change, two additional fields of study stand out: cultural cognition theory and the moral psychology of climate change.

The central argument of cultural cognition theory (Kahan, Jenkins-Smith, and Braman 2011; Kahan 2012) is that people make decisions in line with preexisting cognitive structures and deeply held cultural values. As Dan Kahan, Hank Jenkins-Smith, and Donald Braman (2011) point out, climate change information does not have a blank slate to start from, but encounters cognitive structures and meaning systems that have developed over a long period of time and are difficult to change. Depending on the fit between the new information and the existing belief and value structures, an individual can reject the information or adjust the given belief system to integrate a new idea. In short, worldviews and ideologies are culturally informed lenses that inform and constrain cognitive responses to climate change (Homer-Dixon et al. 2014).

The observed link between climate change skepticism and political conservatism, especially in the United States (Jacques 2012), nicely illustrates both cultural cognition theory and Norgaard's theory of socially organized denial. Climate change, if taken seriously, threatens deeply held conservative values, such as individualism, private property and free enterprise, small government, and antimultilateralism (McCright and Dunlap 2000; Jamison 2010). Hence conservative cognitive structures resist the change required to embrace climate change and support policies to fight it. The cognitive dissonance caused by the scientific consensus on anthropogenic climate change is resolved with climate change skepticism—a cognitive-affective coping mechanism combined with political tactics (Jacques, Dunlap, and Freeman 2008) to protect a coherent conservative ideology.

Whether and how individuals perceive of climate change as a moral problem—the subject of research on moral psychology (Markowitz 2012; Markowitz and Shariff 2012)—determines whether evoking moral arguments might help mobilize action (Skitka, Bauman, and Sargis 2005). In 2015, Pope Francis's involvement in climate change politics through the convening of several high-visibility events and a papal encyclical on climate change demonstrated the relevance of this issue in a religious context. The pope's interventions in climate change politics clearly elevated the moral dimensions of problem, for example, stressing our obligations toward one another and toward the planet to protect the environment as the source of life. Francis introduced a novel moral frame, suggesting that action on climate change will help and might be necessary to improve the situation of the world's poor—a key moral concern of the Catholic Church. His actions raised a number of challenging questions about the relationship between faith, science, and politics and the power of religious belief systems. What is the source of the Pope's power to influence climate change politics? Was he right to use this power? Was he successful? If so, why—and, if not, why not?

More generally, the field of climate ethics (Page 2013) addresses fundamental and often unprecedented moral questions related to climate change, for example, the issue of intergenerational equity or, in Stephen Gardiner's terms (2010), intergenerational "buck-passing." Some experimental research explores the role of fairness in people's beliefs about burden sharing in efforts to mitigate climate change (Gampfer 2014) or the link between moral principles and preferences for certain mitigation policies (Sacchi et al. 2014). As I will discuss in greater detail in chapters 5 and 6, moral reasoning shapes the views of those engaged in global climate change negotiations in important ways.

An exciting subfield has emerged at the intersection between cognitive science and political psychology: the exploration of cognition and emotion in moral political reasoning. Since morality and ideology are often embedded in each other, the question arises whether conservatives respond differently to climate change than liberals do. Moral foundations theory argues that they do. Distinguishing between different moral emotions, moral foundations scholars have demonstrated that liberals and conservatives have different moral-emotional profiles (Haidt 2003, 2013). For example, conservatives emphasize the moral values of maintaining purity and authority rather than preventing harm and caring, which are central moral values of liberals. Consequently, framing environmental policy measures either in terms of purity of or care for the environment elicits very different responses from members of these two ideological groups (Feinberg and Willer 2013). Similarly, Dena Gromet, Howard Kunreuther, and Richard Larrick (2013) demonstrated that ideology influences energy-efficiency attitudes and choices. Much of this research confirms the importance of framing effects in public and policy discourses (Tanner, Medin, and Iliev 2008; Bischof 2010), including the possibility of using different policy frames to motivate mitigation behavior among climate change skeptics (Bain et al. 2012).

This literature also has to grapple with the role of emotions, in particular, moral emotions, in people's responses to climate change. Sabine Roeser (2010, 2011, 2012) argues that emotions are a crucial component of risk perceptions and ethical deliberation on climate change. Other researchers have begun to explore the effect of specific emotions, for example, fear and hope, in climate change messaging (e.g., Stern 2012).

All of these research streams—on belief formation, the psychology and social organization of denial, cultural cognition, and moral psychology—are concerned with the responses of individual voters, public opinion, or sociopolitical movements. Thomas Lowe and Irene Lorenzoni's work (2007) on expert views is an exception to this general focus on the public rather than elites or specific decision-maker groups. So far, there have been no studies applying these rich theoretical insights to diplomats, policy makers, or other actors engaged in the international negotiation process. Given the professional-cultural context of individuals engaged in the climate change negotiations, one would expect cognitive-affective coping mechanisms to be less important for this particular group than the constraints of cultural cognition and ideology. A diplomat assigned to the climate change desk is unlikely to respond to this task with climate change denial. It is more likely that the Habermasian "lifeworld" of climate change negotiators (Depledge 2006, 10) would reinforce certain ideas regarding the purpose and justification of UNFCCC negotiations. Further, negotiators might think very differently about climate change than members of the public, for example, focusing on concepts such as the national interest, power, and vulnerability in a multilateral setting.

Mindmade Politics integrates insights from the various research programs outlined above (e.g., what people know and believe about climate change, how their beliefs change, or what role morality and emotions play), and applies these ideas to a group of decision makers with particular relevance for the success of global efforts to address climate change: those engaged in global climate change negotiations.

Climate Change: A Unique Cognitive Challenge?

The problem of climate change displays a set of extraordinary characteristics that could influence people's cognitive and consequently political responses in possibly unique ways. Several versions of this "climate change is special" argument exist. For example, some authors have suggested that climate change is a "wicked" (Rittel and Webber 1973; Prins et al. 2010) or even "superwicked" problem (Levin et al. 2009, 2012), which implies that it is harder to solve than other political problems and that straightforward technical solutions might not be available. Collaborative solutions to climate change might therefore require a political or governance approach that is altogether different from the one the international community has been using (Prins et al. 2010). We might even need a different kind of science (Verweij et al. 2006). Even though he classifies mitigating climate change as an aggregate-effort problem (i.e., the more countries contribute, the more relief they can achieve), Scott Barrett (2007, 9) argues that "global climate change ... is almost certainly the hardest [problem] for the world to address." But, so far, none of these definitions has been analyzed from a cognitive perspective. How does the mind deal with climate change's special problem characteristics?

Although the concepts of and criteria for "wickedness" or "superwickedness" remain contested, the notion that there might be something special, something particularly difficult, about the climate change problem, is intriguing. In the sections below, I outline a set of characteristics that might make climate change a possibly unique political-cognitive problem, and I select and combine some of the characteristics that might present particularly problematic obstacles to international cooperation.

Special Problem Characteristics

Pervasiveness   Climate change is an unusually pervasive problem. By that I mean that, beyond its global nature, which it shares with other world problems, climate change has causes and effects at almost every imaginable natural and social system level. Further, there are multiple, often surprising, cross-level linkages and dynamics. Pervasiveness introduces unprecedented, daunting degrees of complexity because causes and effects of the problem show up everywhere and anytime, leaving no sphere of social life untouched and requiring solutions far beyond technology and politics. For example, greenhouse gas emissions and land-use change are not limited to a few countries or to a few industry players, as in the case of ozone-depleting substances, such as chlorofluorocarbons (CFCs), but include individuals, households, firms, industries, transnational networks, and governments around the world. Consequently, addressing climate change requires almost universal changes in behavioral patterns, institutions and infrastructure, especially with regard to energy use. The impacts of climate change can be experienced by ecosystems and regional climate systems, by microorganisms, plants, and animals, indeed, by all life on the planet from the depths of the acidifying oceans to the heights of the warming atmosphere. Climate change affects all natural systems that sustain human societies, that provide food, clean air and water, and other natural resources. The problem fundamentally challenges not only all current economic structures, but also all social institutions, including those providing health care, education, and public services.

Another contrast between climate change and other global governance challenges is the fundamental importance of implicated structures for human civilization, progress, and productivity. Energy production and supply systems are the backbone and lifeblood of any society, which makes efforts to change, much less do away with, the existing constellation of actors, resources, and infrastructure extraordinarily difficult and qualitatively different from, for example, replacing in a handful of industries chemicals that deplete ozone with ones that do not. Transforming the global fossil fuel industry will be disruptive not only for the industry itself but for entire societies that depend on it. The economic, political, and social changes required are staggering.

Pervasiveness and the need to transform global energy systems, in themselves, make climate change a special collective-action problem. Global challenges that require many actors to collaborate are not uncommon. Climate change is one such challenge since somewhere between ten and twenty states are needed to collaborate to significantly reduce global greenhouse gas emissions and manage land-use change (Victor 2006, 95). But, beyond this minimum requirement of multilateralism, many more actors—and not only state actors—have to collaborate to set the world on a path to carbon neutrality. Developing countries might have to use certain energy technologies, but not others; companies might have to change their business models and surrender to certain types of regulation and taxation; individuals might have to accept limits to their personal freedom—the freedom to eat meat, drive, or fly whenever they want, for example. All of this takes place in a global environment where power and vulnerability with respect to climate change are very unevenly distributed, and the collective interest is contested, where inequality within and across countries is high, the global population keeps growing, urbanization is increasing, and many global governance institutions are showing signs of stress. Taking these circumstances into consideration, the collective-action problem in response to climate change appears to be qualitatively different from—indeed, unimaginably harder than—getting twenty states to agree to reduce GHG emissions.

Uncertainties Global climate governance needs to deal with greater uncertainties regarding numerous important questions about both natural and social systems than those facing any other global governance regime. These uncertainties include the various responses of the climate system to increased concentrations of greenhouse gases in the atmosphere, the impacts of climate change–induced environmental changes on human well-being and institutional functionality, and, finally, the various effects of political and policy decisions on both the climate system and societies over long periods of time. Again, uncertainties are not unique to climate change (Levin et al. 2009), but the scale and dimensions of those related to global climate governance might well be unprecedented.[2]

Long Time Scales Some of these uncertainties are linked to the circumstance that the global climate system works on extremely long time scales and displays significant time lags between an initial change in the environment and the system's response, for example, the increase of greenhouse gas concentrations in the atmosphere and the associated increase of global average surface temperature (Clark et al. 2016). Climatic changes take place over decades, centuries, and millennia, rather than just years or election cycles. Similarly, there is potentially a large time lag between the implementation of a climate change policy and the policy's impact on the climate system. These lags make it difficult to predict with precision when or over what time period the effects of such policies would manifest.

Under these conditions, it becomes increasingly difficult, if not impossible, to perform a standard cost-benefit analysis with temporal discounting. Indeed, the combined challenge of long time scales and major uncertainties has given rise to a debate among distinguished economists about the general utility of such analytic tools in the case of climate change. The arguments center on the possibility of "fat-tail" events—low-probability but high-impact occurrences with catastrophic implications for a society (Weitzman 2009; Nordhaus 2012). Derek Lemoine and Christian Traeger (2012) are among the first to model the economic impact of climate tipping points and their interactions with climate change policy; Timothy Lenton and Juan-Carlos Ciscar (2012) have integrated climate tipping points into their assessment models to offer more realistic decision support, bridging natural and economic systems. But these ideas are far from becoming a part of mainstream economic thinking. And even if economic frameworks eventually address the possibility of catastrophic change, temporal discounting is standard practice not only among policy makers but also among the public (Jacobs and Matthews 2012), where it is likely to guide behavior indefinitely.

2. For details regarding the treatment of uncertainty in the fourth Assessment Report of the IPCC, see the "Uncertainty Guidance Note": http://www.ipcc.ch/meetings/ar4-workshops-express-meetings/uncertainty-guidance-note.pdf.

Finally, the deep time dimensions of climate change raise a number of important questions regarding intergenerational ethics, for which there are so far few, if any, satisfactory analytical tools and answers (Gardiner 2011).

Limited Observability Not much is known yet about the relationship between experience and beliefs, but it clearly plays an important role in shaping responses to climate change. The direct, personal, sensory experience of a phenomenon, such as the flooding of your own house, will leave a much stronger mark on your mind than reading in the newspapers about other people's houses flooding (Dessai et al. 2004). It might even make a difference whether these other flood victims have houses near yours, share your nationality, live in a country you have previously visited or in one you know nothing about. Depending on how much you care about these people, you might also care about the likelihood that their plight might be linked to climate change.

That leads to another puzzle concerning the experience-belief relationship: what causes what? Initially, researchers believed that the experience of a climate change–related extreme weather event increased the likelihood of believing in the reality of climate change and hence increased the support for climate change policy. But a prior conviction about the reality of climate change might actually influence whether someone believes a certain weather event to be related to, if not actually induced by, climate change. As Teresa Myers and colleagues (2013) demonstrate, the process of motivated reasoning—"seeing" climate change because one believes in its existence—is as real as the belief-reinforcing effect of climate change–related events.

But the picture is even more complicated than that. The climate change problem is a moving target from a human and especially a political perspective. Our senses provide us with severely limited abilities to observe or directly experience the threat and full consequences of climate change. Some impacts can be partly observed. For example, you might be able to observe some gradual sea-level rise in a particular place in the course of your lifetime, but global sea-level rise takes place over several centuries and has different effects along the various coastlines of the world (Hansen et al. 2015). Similar arguments can be made about ice-sheet loss and ocean acidification. Any individual might get a glimpse, but nobody can get the full picture. Some impacts, like the increasing frequency and severity of extreme weather events, might be easier to detect than others, such as ocean acidification and the extinction of species. Indirect observations, for example, via reports of other people or the media, photographs, or scientific writing, might help you get a better picture, but they do not have the same cognitive effect as direct experience.

The problem of limited observability is particularly important when it comes to causal relationships: our ability to correctly identify the link between causes and their effects. We simply cannot see, feel, or otherwise experience how much warming is caused by the emission of one metric ton of carbon dioxide, or by one commute to work in the family SUV. Accepting that our senses are of limited use when it comes to climate change, all our decision making in that regard has to rely to a large extent on abstract and synthesized information provided by scientists and technical experts—and on our imagination. The critical importance of scientific experts and public communication processes for climate change politics matters in many ways. First of all, it means that climate governance depends on indirect observation rather than direct experience with important cognitive implications, as mentioned above. One of these cognitive implications concerns the role of emotions and affect. The more we have to use abstract and technical information when addressing a problem, the less we are able to access the full spectrum of cognitive information our minds normally use in decision making. Devoid of emotional information, the abstract and scientific basis of our climate change conversations limits the range of responses to the problem that would otherwise be available to us.

These observations need to be qualified in two respects. First, many individuals, especially in the developing world, are already experiencing diverse effects of climate change in their daily lives. Less reliable weather patterns, more frequent and more intense extreme weather events, and sea-level rise are often associated with property damage, loss of life, and rising insurance costs. Individuals with knowledge of climate change are likely to associate these personal experiences with climate change. They are also more inclined to accept indirect reports of others about similar experiences and losses as instances of climate change impacts. The same phenomenon of experience-based beliefs about climate change has recently appeared in the developed world, with increasing floods in the United Kingdom, wildfires in Australia, and sustained drought and intense East Coast storms in the United States.

Climate Tipping Points As defined by Timothy Lenton and colleagues (2008, 1786), a tipping point is "a critical threshold at which a tiny perturbation can qualitatively alter the state or development of a system." Over the last few years, climate scientists have pointed out that the global climate system and various regional climate subsystems could exhibit tipping points, signaled by changes in, for example, various ice sheets (large ice volume vs. none), the Atlantic thermohaline circulation (on vs. off), and the Indian summer monsoon (strong vs. weak). Researchers are cautious, however, about the possibility of global-scale climate tipping points (Lenton and Williams 2013) and continue to emphasize major uncertainties about the conditions for and timing of such events (Lenton 2012).

Although Wallace Broecker (1987) was among the first to express concern about the possibility of climate tipping points, the tipping point concept is not new or specific to the global climate system. Tipping point is often used interchangeably with the terms threshold, regime shift, or critical transition. And not only does it apply to change in natural systems like the climate system, but it can also frame our understanding of change in social systems. Indeed, it was popularized by Malcolm Gladwell, who used tipping points to refer to phenomena such as the spread of a fashion fad in his 2002 book *The Tipping Point*. Researchers increasingly think about transformations of societies toward sustainability or zero-carbon economies in terms of desirable tipping points (Westley et al. 2011). I adopt Broecker's and Lenton's definitions of tipping point, which emphasize the nonlinear or rapid character of systemic change when such a point is reached (Broecker 1987; Lenton et al. 2008). Internal system processes such as feedback effects then drive this change.

Tipping as a specific system behavior could have very serious implications for the well-being of human societies, which have evolved in and adapted to the stable global climate system over the last 10,000 years. The possibility of climate tipping points therefore poses major challenges for the design of climate governance institutions. Gardiner (2009, 140) suggests that the growing awareness of the possibility of tipping points should be welcomed: by overcoming the current political inertia, it could "help us to act." Mark Nuttall (2012, 97) argues that the major discursive power of the tipping point idea prompts "discussion characterized by a nervous anticipation of the future," whereas Mike Hulme and Rob Bellamy (2011) contend that different value systems determine the effect of tipping point concerns on individual beliefs about climate change risk and action. So far, however, we do not know whether the tipping point concept affects the beliefs and decisions of climate change negotiators and, if it does, how.

Multiple Simultaneous Stresses   An often-underestimated feature of climate change is its likelihood to produce multiple simultaneous stresses on social systems, rather than onetime, limited emergency events. Thus climate change–induced droughts, fires, plagues, floods, and coastal storm surges in different regions of the world could place multiple simultaneous stresses on global food systems that, in turn, could trigger crisis cascades in national social and economic systems. These systemic stresses could also have repercussions for health systems, international trade, poverty alleviation, disaster risk management, and international security (Homer-Dixon et al. 2015). One such case occurred in 2010, although it remains unclear whether all its individual stressors can indeed be linked to climate change: a heat wave and fires in Russia, floods in Pakistan and China, and a drought in China all converged to produce major stresses on the global food system, which might have contributed to the 2010–2012 uprisings in North Africa and the Middle East (Werrell, Femia, and Slaughter 2012).[3]

Lack of Intentionality and Moral Rules   Finally, two characteristics of the climate change problem are particularly interesting in the context of conflict and security. First, the lack of intentionality or hostility on the part of those who cause climate change–related harm (indirectly with greenhouse gas emissions and land-use change) and the vast geographic distribution of climate change contributors make it very difficult to attribute blame for any specific climate change–related event or phenomenon to a certain group—a state, a company—or even a group of individuals. In other words, it is almost impossible to identify an enemy or somebody to blame. This

3. See Tom Gjelten, "The Impact of Rising Food Prices on Arab Unrest," NPR *Morning Edition*, February 18, 2011, http://www.npr.org/2011/02/18/133852810/the -impact-of-rising-food-prices-on-arab-unrest.

characteristic also renders strategic thinking in the conventional sense of international relations—game-theoretic responses to enemies in an anarchic world—useless. And, second, the lack of moral rules about behavior resulting in atmospheric or environmental changes implies a lack of moral and emotional arousal (Grasso 2013), often a key ingredient for political action against a looming threat.

Cognitive Responses to Special Problem Characteristics

Some or all of these characteristics of the climate change problem might affect the cognitive ability of individuals and the international community to respond to the challenge effectively by using the tools of international diplomacy. In the following subsections, I describe what these cognitive effects might be and explain how they could influence decision making in the UNFCCC negotiations.

Pervasiveness and the Loss of Hope Preparedness to act on climate change requires, first of all, the recognition or belief that success is possible. Such a belief would include some definition of success, which is translated into a set of goals, such as limiting global warming to 2°C, and pathways toward those goals, such as mitigation timetables and energy system transition planning. Selecting and pursuing goals is a form of agency, in this case, the collective agency of the international community. When it comes to climate change, success is dependent on the actions of other states, and each actor's corresponding assessment of others' preparedness to make and live up to a commitment. This assessment can strongly affect the individual actor's sense of agency, and of course, the chances of coming to a cooperative agreement. Perceiving others to be unwilling or reluctant to contribute to collective goal achievement can easily lead to resignation and hopelessness.

More generally, goal pursuit, pathway thinking, and agency are connected by the cognitive-emotional phenomenon of hope (Snyder 2002). Based on the psychological theory of hope, an actor's loss of hope when linked to perceptions of others and their intentions can diminish the actor's sense of agency (McGeer 2004).

Success, as it is currently defined in the global policy process, necessitates the cooperation of many actors in the form of an effective agreement under the UNFCCC umbrella.[4] Given the unprecedented scale and complexity of the climate change problem, past negotiation failures, and prevalent perceptions that major emitters are reluctant to act, one would expect that some actors experience hopelessness at certain points in the negotiation process. Loss of both hope and any sense of agency, and especially the dynamics of spreading hopelessness, could play an important role for the collective ability to establish and implement a cooperative agreement like the Paris Agreement.

Treating the loss of hope as a potential cognitive response to the pervasiveness of the climate change problem, I was curious whether and when UNFCCC negotiators experienced hope and hopelessness in the course of their work, and how hopelessness affected their negotiation behavior.

Time Scales, Limited Observability, and Myopia Most personal, ethical, and political decisions in modern societies have a limited time horizon. For an individual, this can range from days to a few years, in rare cases a few decades, but the future "goes dark" for most people around fifteen to twenty years from their present (Tonn, Hemrick, and Conrad 2006). Most political decisions in democracies are driven by election cycles (about five years) and economic dynamics, fewer by the lifetime of infrastructure projects (several decades) or major social security systems (a generation). In global governance, a fifteen-year time horizon is surprisingly common, for example, in the context of establishing global development goals. Decision making with a time horizon of more than fifty years is practically unknown in the public sector.

These familiar time horizons imply that there is a fundamental mismatch between those for most decision making by modern societies and political actors (years to decades), and those for most aspects of climate change (decades, centuries, and millennia). It also means that humans have no experience to build on when making decisions on climate change.

The power of political myopia can best be seen in the prevailing practice of temporal discounting in economic and political decision making. The standard rationalist approach to political choice and policy making is cost-benefit analysis—comparing different available paths of action,

---

4. For some actors, that assessment might change with the development of geo-engineering technologies and a growing understanding of their costs and effects.

one always chooses the option that maximizes the difference between expected gains and losses (net benefits) over a distinct time period. Economists have long debated the role of time in the process of cost-benefit analysis: how does one value costs and benefits occurring in the present compared to those expected to occur in the future? Two distinct responses to this question have been developed (Loewenstein and Elster 1992). One response says that it is "natural" for individuals to assign a lower value to future costs or benefits, pointing to a number of reasons why humans are inclined to place a higher value on the more certain promises of the "many temptations of the present" than on the less certain promises of the future. A second response says that present and future assets should be treated equally.

So far, the voices in favor of temporal discounting have won the argument. A study by Alan Jacobs and Paul Matthews (2012) offers empirical evidence that not only policy makers but also citizens practice temporal discounting. This is of particular relevance for decision making on climate change because effective mitigation policies would have considerable and well-understood short-term costs, but more uncertain long-term benefits. In other words, the current generation is required to pay a relatively certain price for preventing future generations' less certain harm. Under these conditions, a CBA typically results in inaction—the perceived costs simply outweigh the perceived benefits, as long as the current generation of decision makers sufficiently discounts the future.

On the other hand, in the global climate change debate, not all actors seem to succumb to myopic decision making in the face of this long-term problem. Some appear to take long-term consequences and time lags seriously into consideration, and, having done so, they strongly argue for prompt political action despite its short-term costs. What causes the difference between these perspectives? A rational-choice scholar would suggest that the long-term thinkers probably have more to lose in the future, or that, being poor, they would not incur any the costs of the prompt action they are calling for.

Other explanations for the different attitudes observed among climate change negotiators could include differences either in their social value systems and definitions of intergenerational justice or in their attitudes toward nature and environmental stability. Both of these factors are elements of a society's social identity and its dominant ideologies. Finally, some individuals might simply have a higher capacity for abstract thinking and imagination than others.

Using this mismatch of time horizons for decision making and the climate change problem as a point of departure, I analyze the different cognitive-affective processes for dealing with the long time horizons of climate change, in particular, the different emotions associated with sensory experience compared to abstract scientific information. I also take a brief look at imagination. How do individuals imagine the long-term future, in particular climate tipping point events and their social consequences?

Limits of a Cognitive Approach

Like any analytic lens, a cognitive approach has not only certain strengths that help us focus on and understand parts of reality, but also clear limitations. A cognitive analysis draws attention to a specific set of variables that are grounded in mental activity. These variables include thoughts, emotions, beliefs, and belief systems. They raise questions concerning their origin and development over time, the ways in which they might be shared in a group, and the ways in which they might affect social and political processes.

What a cognitive analysis is not able to do is to capture or speak to the relevance of other variables, especially the material and behavioral variables that much of political science is concerned with. In contrast, my conception of cognition as both a multilevel and a relational process—the mind as embedded in a multilevel system, connected to other minds as well as to the material world of objects and beings—provides multiple opportunities to link the cognitive analytic framework with material or other ideational frameworks in international relations scholarship. For example, a cognitive analysis proceeding from this conception would suggest that decision makers' beliefs about the vulnerability of their home countries to climate change–induced storms and floods and their related negotiation position that a group of developed countries should provide financial support for adaptation measures are rooted in their mental assessments of the material conditions of the international system (i.e., the distribution of wealth, power, and climate impacts among states) and in their ideas about the appropriate norms of behavior that apply in this situation. Those two variables—the structure of the international system and the existence and relevance of norms—are the subject of structural and constructivist theories. They can easily be connected with and speak to the insights generated with a cognitive analysis.

More generally, one could criticize a cognitive research program for implicitly reducing complex sociopolitical phenomena to individual mental processes, especially if one rejects the idea that all social phenomena have their origins in individual minds. Mary Douglas (1986) argues that "the social" is prior to any individual cognitive processes and in fact determines individual thought. From her perspective, by confusing independent and dependent variables, a focus on individual minds fails to acknowledge the priority of the social processes, conditions, and institutions that create and result in individual beliefs.

Again, in contrast, the cognitive theories I adopt and build on in *Mindmade Politics* do not consider the social world to be secondary, much less irrelevant, to individual cognition. Rather, they conceive of the relationship between the social and the individual as interdependent and mutually affecting each other. Although the social elements might serve both as strong constraints and as enablers of individual thought, the individual mind also affects and shapes what is socially shared. My conception of cognition goes even further, emphasizing a role not just for social factors in cognition, but also for the nonsocial, material, and natural environment.

None of these three elements—the individual mind, social conditions, or the natural environment—should be prioritized. Rather than focusing on any one of them, I contend that the interaction and mutual interdependence between these elements are key to understanding existing beliefs and belief dynamics.

It is unclear how much cognitive "freedom" or self-determination rests with the individual and to what extent an individual's beliefs have been received from the social environment or even imposed by it. But, without allowing for individual and cognitive agency, it would be hard to explain individual differences, diversity, and idiosyncrasies in existing belief systems. If all individual thought was predetermined by the social, from what would novelty arise? And how would the specific social conditions come to be in the first place? Even when acknowledging the crucial interactions between the individual, the social, and the environmental, much can be gained from using the individual mind as an entry point to the analysis.

The rules that apply to the individual mind constrain what types of belief systems are possible, and those constraints also apply to shared or collective beliefs.

A Cognitive Framework for Political Analysis

Advances in the cognitive sciences over the last three decades have resulted in theories and insights about the mind, thought, and decision making that are not only fascinating, but also accessible and that may play a highly constructive role in the research of other disciplines. They are particularly alluring for political scientists and international relations scholars, who seek to understand how and why political decisions come about. Despite early interdisciplinary efforts and a particular interest in the role of emotions in international political life, research projects into political decision making have remained limited in scope and are constrained by the lack of appropriate methodologies to identify and track thoughts and emotions.

This chapter's quick journey through cognitive theory and the insights of chapter 2 enable us to establish the basic features of a cognitive analytic framework with a focus on global climate change politics. They must include the following.

Cognitive Elements

Each actor has specific mental representations—ideas, beliefs, concepts—concerning the nature of climate change as a subject of international diplomacy. We can look to international relations scholarship to find some clues about the substance of these representations. Based on the dominant schools of thought in international relations, we would expect political actors concerned with climate change to hold beliefs about (1) the costs and benefits associated with climate change itself and with the policy responses to climate change; (2) the identity of the self and others (in- and out-groups); and (3) justice and applicable norms with regard to climate change. We would also expect all these political actors to have mental representations of the climate change problem itself, including its special characteristics, such as pervasiveness, uncertainties, and the possibility of climate tipping points. These mental representations would be rooted in climate science and sometimes in personal experience.

Emotions
Inseparable from concepts and thoughts, emotions are important factors in risk perceptions and moral judgments. Identifying dominant emotions and differences in the emotional profiles of the belief systems of various actors is a key task of a cognitive analysis. As I hypothesized above, hope and hopelessness as well as fear and depression might be important emotions to consider.

Cognitive Structures
Cognitive elements (mental representations) and their emotional valences are connected with one another, forming networks that have distinct structural features. We could call these structures a cognitive architecture, and we would expect this architecture to be emotionally coherent and to be relatively stable over time.

Individual versus Collective Cognition
It is important to distinguish between individual and collective cognition. The links and interdependencies between these two levels of analysis present some of the most interesting theoretical and empirical challenges of cognitive science. Currently, this relationship is best understood in terms of multilevel interacting mechanisms, involving cognitive mechanisms at the individual level and social mechanisms at the group level. Each person participating in this research project had a set of specific, idiosyncratic beliefs, but, at the same time, was also a member of one or more groups that shared certain ideas and convictions. It is this ability to have collective beliefs that allows individuals to be functional actors in the international system, to represent entire countries, for example, or to form alliances in the climate change negotiations. The details of this relationship between individual and collective cognition, including how they influence each other over time, remain to be uncovered.

Cognition and Material-Social Realities
Recognizing that cognition is not a mind-only activity, isolated and independent of factors external to the mind, is a fundamental prerequisite for undertaking a cognitive analysis. Most mental representations refer to things in the material-social world, whether physical entities, like trees or cars, or socially constructed ones, like states or wars, which depend on

shared beliefs about the existence and meaning of those entities among many people. More generally, mental representations tend to relate to tangible objects or to processes and phenomena we can observe or experience with our physical senses, such as artifacts, other people, written and spoken words, and the natural environment. As a relational process, cognition connects mental activities to material-social realities and is best understood as a set of causally intermediate processes linking material-social factors, on the one hand, and the decisions and behaviors of individuals and groups, on the other.

Cognitive Processes

Creating and maintaining emotional coherence is one of the most important rules of mental processing. The theory of emotional coherence assumes that cognitive processes have emotional valences and that decisions are based on a process of multiple constraint satisfaction, taking into consideration both the cognitive acceptability of a mental representation and its emotional valence.

Rational choice, the effort to maximize utility by calculating costs and benefits of certain decision options, is another relevant decision-making process in international politics. I treat rationality as a cognitive process. In the narrow sense of rational-choice theory, rational decision making is likely to be severely constrained with regard to climate change because of high levels of uncertainty, difficulties of quantifying certain costs and benefits, and long time horizons. I suspect that, despite these significant challenges, the political actors involved in climate change policy making or negotiations attempt to be rational, using familiar categories (costs and benefits) and familiar time frames (years and decades). Long-term futures are not likely to figure in their thinking.

More generally, rather than assuming that the rational-choice analyst can replicate the rational decisions of a policy maker or negotiator, cognitive analysis seeks to describe the actual cognitive processes underlying the observed decisions and behavior of either. Cognitive analysis should reveal whether and to what extent policy decisions or negotiation positions are based on a rational cost-benefit analysis, or on assessments of morally appropriate behavior and the relationship between identity and norms. Did US President George W. Bush calculate the economic cost of a carbon tax when he refused to submit the Kyoto Protocol to the Senate for ratification?

Did European Union policy makers push ahead with cap and trade because they saw themselves as climate and technology champions and simply disregarded the calculated economic costs? Or did the US president and the Europeans simply have different definitions of economic costs in the context of climate change? Providing answers to these questions, cognitive analysis can verify the cognitive assumptions of major international relations theories and perhaps point to gaps in these theories if they do not fully capture the observed thought processes.

These six features of cognitive analysis form the framework for the empirical research that will be presented in chapters 5 and 6. By integrating cognitive theory, international relations theory, and some features of climate change science, they create an analytic lens for studying global climate change politics.

4 Methods and Tools for Cognitive Social Science

Engaging in cognitive social science research is methodologically challenging. So far, there is a rather limited set of tools to identify and observe changes in beliefs and belief systems, mental structures and emotions. Neuroscientific methods such as functional magnetic resonance imaging (fMRI) cannot provide answers to key social scientific questions concerning the content of beliefs or the cognitive-affective processes involved in decision making. But, with the help of computer models using behavioral and other data gathered from surveys, focus groups, interviews, and experiments, theoretical cognitive scientists are beginning to develop tools that can.

For this research project, I combined two methods, one qualitative and one quantitative, to explore the conceptual and emotional content of the belief systems of individual study participants and to identify shared belief systems among groups of participants. This chapter introduces the methods—cognitive-affective mapping (based on semistructured interviews) and the Q method—and outlines their key features, their strengths and weaknesses, and the powerful effects of combining them. It also presents the rationale for working with those engaged in UNFCCC negotiations, in particular, diplomats, and it provides a detailed description of the process for selecting study participants from among them.

Participant Selection

A key design feature of this research is the engagement of individuals who partake in the global climate change negotiations, especially diplomats representing states that are parties to the UNFCCC, but also representatives from a variety of NGOs and private sector organizations (e.g., industry associations and companies)—the so-called observer organizations. The decision to work primarily with diplomats raises a number of important methodological questions.

Investigating the private beliefs of diplomats rather than those of their respective political masters within their national governments has major advantages, but also some serious drawbacks that place important limitations on the conclusions that can be drawn. A central issue is the fundamental distinction between private beliefs and public negotiation positions (Feldman 1988; Hamm, Miller, and Ling 1992; Niemeyer 2011). Private beliefs are specific to an individual and generally not revealed or considered relevant in international negotiations. Negotiation positions are sets of ideas that are shared within particular groups of diplomats (national delegations) as well as with the diplomats' political principals in their respective national capitals. Negotiation positions reflect mandates diplomats are both tasked to fulfill and constrained by. These positions are frequently shared with other negotiation parties and the press. That said, one has to ask whether there is any utility in exploring the private beliefs of individual negotiators; after all, they could be considered as mere messengers of their domestic political masters, whose mandates they are powerless to alter in the creation of cooperative agreements between states. Below, I offer three arguments why working with diplomats was both useful and necessary for the purpose of this project and why it offered unique advantages over working with other potential study participant groups.

The private beliefs of individual climate change negotiators contain the *most detailed, richest points of view* that exist regarding the global aspects of the climate change problem. They often differ from the private beliefs of domestic actors due to the negotiators' unique experience of climate change as a global governance challenge. It is the professional responsibility of diplomats to address climate change in a multilateral setting. They have maximum access to the changing scientific information about climate change, they are frequently exposed to the views of other global actors, and they have to continuously present and justify their national position in a coherent manner. This constant exposure to other actors' views uniquely expands the perspectives of diplomats—the range of ideas and arguments they have to consider and respond to. At the same time, it can also lead to narrow understandings of the climate change problem, to perceiving it purely in terms of multilateral treaties—divorced from on-the-ground realities and national contexts.

Diplomats are also the links between domestic and international political processes, making their cognitive realities crucial for cross-scale interactions. They are two-way communication channels, relaying domestic messages to the international community and carrying decisions, questions, and tasks from the multilateral forum into their respective domestic political spheres. Because of these specific circumstances, the private beliefs of climate change negotiators should be more comprehensive and sensitive to global complexities than those of domestic political actors, who are not required to take all these circumstances into account.

Further, diplomats focus on the multilateral context, which is the subject of this research, rather than on the domestic politics of climate change. The global political debate on climate change is different from the domestic ones; it has not only different participants, but also different conceptual elements, processes, and technicalities. In short, global-level discourses on climate change and cooperation differ from domestic discourses, although naturally there will be major overlaps between them.

What is more important, the private belief systems of climate change negotiators, in particular diplomats, contain what one could call the "current possibility space of problem definitions and solutions." In other words, this research identifies belief systems—viewpoints—regarding what is possible, whether or not these viewpoints are publicly revealed in the form of negotiation positions. The private beliefs of diplomats can differ significantly from the official negotiation positions they present in the UNFCCC context, but these beliefs reflect the valid—coherent and understandable—private viewpoints of persons of the same nationality, viewpoints that are shaped by the same multiplicity of social, political, and cultural factors that influence the diplomats' respective domestic decision makers, but that are also subject to an additional set of global influences. Investigating them thus leads to a more complete understanding of the landscape of possible ways of thinking about and responding to climate change.

Working, in particular, with diplomat negotiators as study participants is also preferable because their belief systems are *comparable* to each other due to the similarity of the individuals' roles as representatives of governments in the UNFCCC process. A study involving domestic decision makers would face major difficulties in this regard. A great many different actors (e.g., members of various ministries, departments, or agencies, industry players, parliamentarians) are involved in the process of determining negotiation positions or national climate change policies. The relevant sets of actors differ across countries, making a useful comparison difficult, if not impossible. Further, there are major differences between the domestic discursive landscapes of different states, impeding a successful Q study design (see below).

What is more, in the case of small island developing states (SIDS), delegates often have significant freedom to determine their respective countries' negotiation positions and climate change responses strategies more generally. Their private beliefs tend to overlap strongly, if not completely, with what they argue or propose publicly in the formal negotiation process.

That said, the goal of *Mindmade Politics* is not to improve the clearly limited power of negotiators to create a cooperative agreement, but, rather, to understand the nature of existing beliefs of climate change negotiators and the relevance of these beliefs for multilateral cooperation. For this purpose, I assume that these beliefs and the viewpoints they form would be valid in political debates and that there is a possibility of engaging domestic political audiences based on them.

It is worth noting that my decision to work with those engaged in global climate change negotiations, in particular, diplomatic representatives of the negotiating states, has major implications for the kinds of belief systems this study is likely to uncover. Most of these individuals were part of a global elite whose high level of education and professional training affected their worldviews. One could expect that the similarity of their professional backgrounds might lead to similarities in their views concerning multilateral negotiations and climate change, for example, creating an undue focus on the UNFCCC and diplomacy as solutions to the problem. But, as the empirical findings in chapters 5 and 6 demonstrate, concerns about excessive similarity in private beliefs within the sampled population were unfounded. Although there was a natural focus on the multilateral process, study participants had a very diverse set of beliefs that involved climate governance actors and spaces far beyond the UNFCCC. They also displayed diverse degrees of familiarity with climate change impacts on the ground and the related concerns of ordinary citizens. One factor contributing to this diversity in beliefs was the mix of participants: although two-thirds were diplomats; the remaining third were representatives of so-called observer organizations, including NGOs and corporations.

In summary, studying the private beliefs of diplomats serving as climate change negotiators is valuable for three reasons. First, these beliefs are the most relevant but also the most specific and detailed with respect to the subject of this study: the global aspects of climate governance. Second, they contain the current possibility space of solutions to the climate change problem, as well as potential lessons about the nature of cognition. And, third, they lend themselves to a rigorous cross-national comparison that would not be possible with the private beliefs of domestic political actors.

The Selection Process

The process of selecting study participants for this research project had to satisfy the requirements of both cognitive-affective mapping and the Q method, which will be discussed in greater detail later in this chapter. The key goal of the selection process was to identify different but comparable points of view held by those engaged in the global politics of climate change; it therefore had to ensure a minimum diversity among the perspectives of study participants without the need to be globally representative or comprehensive.

With respect to diversity, the participant selection requirements for cognitive-affective mapping (CAM) and the Q method are closely aligned. The Q method is somewhat more demanding regarding the strategic sampling of participants, emphasizing inclusion of the most important viewpoints on the topic at hand. But because the aim of this research was to explore the nature of negotiators' existing belief systems and to theorize about their content and structure, not to assess their political relevance or how existing differences between them might be reconciled in the negotiation process, the importance of a particular viewpoint was not a key concern.

Participant selection with the aim of capturing a diversity of views had to simplify the existing complexity of the many variables that could influence a person's views on multilateral cooperation and climate change, while dealing with the practical difficulties of recruiting study participants from among those engaged in an ongoing, contentious, and highly demanding political process.

Diplomats   For diplomats, this simplification could be achieved in various ways. One possible strategy could consist of selecting two countries from each of the major negotiation blocs (e.g., Umbrella Group, AOSIS, G77 and China). I did not choose this approach, however. First, the membership of these groups is continually changing and significantly overlaps from one group to another. Second, the reasons for being in a negotiation group are likely to be political or historical rather than a shared set of interests and views regarding global climate governance. The best example is the G77 and China, whose 134 member countries have diverse and often conflicting views on climate change, but which is held together by a specific historical context (Najam 2005). Third, the privately held beliefs of climate change negotiators that this study seeks to understand might well differ from the positions presented by the negotiation groups they are part of.

Instead of mirroring existing organizational patterns, I identified two variables that could reasonably be expected to influence the perspective of individuals representing a certain country. The first variable was the home country's relative contribution to climate change in terms of its CO_2 emissions; this variable determined both the country's power to contribute to the solution of the climate change problem through mitigation, and the potential costs it might have to bear in doing so. The intuition underlying this variable was simple: citizens in high-emitting countries tended to have more concerns about the costs of action and would therefore be more reluctant to support a cooperative multilateral framework than citizens in low-emitting countries. The second variable was the expected and perceived severity of present and future climate change impacts within the home country, in other words, its perceived vulnerability. Citizens in highly vulnerable countries were likely to focus on the costs of nonaction or delayed action and to favor a cooperative multilateral approach to avoid expected harm and damage.

I distinguished between three national CO_2 emission levels—high, medium, and low—and two levels of vulnerability—high and low. Combining these two variables resulted in a 3×2 matrix (see table 4.1) and consequently six different groups of countries with distinct emission-vulnerability

Table 4.1
Study participant groups—diplomats

		Level of national CO_2 emissions		
		High	Medium	Low
Vulnerability	High	1-HH	2-MH	3-LH
	Low	4-HL	5-ML	6-LL

profiles. Based on these characteristics, I labeled each group of countries with two letters, the first letter representing the CO_2 emission level, the second letter representing vulnerability. For example, group 1, consisting of countries with high CO_2 emission and high vulnerability levels, was labeled "1-HH." Group 2, countries with medium emission and high vulnerability levels, was labeled "2-MH," and so on.

This group matrix provided the starting point for study participant selection. All party countries to the UNFCCC were placed in the matrix. Among all the countries in a particular group, I selected a minimum of three and reached out to individual members of the countries' national delegations, using lists of those participating in COP 17 (Durban, 2011) provided by the UNFCCC Secretariat to recruit study participants, regardless of their roles in their respective delegations. Of the problems associated with using the group matrix in selecting participant diplomats for the study, I address the three most important ones in turn.

Problem 1. The variables I used to create study participant groups (emission and vulnerability levels) had to be based on objectively measurable qualities of the home countries of these individuals. But the candidate participants might have had very different subjective assessments regarding these variables and might therefore have placed their home countries in different groups from the ones I chose. For example, the measures I applied placed Canada in group 4-HL with high emission and low vulnerability levels. Some candidate participants from Canada, however, might have objected and pointed out that the fast-changing Arctic and their country's long coastlines made Canada highly vulnerable to climate change, and therefore it belonged in group 1-HH.

This discrepancy between objective and subjective assessment could have undermined the goal of including a diversity of views because some study participant groups might have been underrepresented, or not represented at all. This seemed an insurmountable problem, given the impossibility of eliciting the views of study participants before actually selecting them. But because this problem applied to individuals in all groups, the effects of inappropriate group assignments canceled one another out across the entire population of study participants.

Problem 2. The distinction of three emission levels and two vulnerability levels implied a major simplification of the complex reality of the processes and system characteristics involved. Determining numerical cutoffs between high, medium, and low emissions was not straightforward given the massive differences among countries' emission profiles. Creating three rather than two categories reflected the following characteristics of these emission profiles. First, there was a small set of countries whose emission levels were one or several orders of magnitude higher than those of all other countries. This group of emitters overlapped significantly with the Group of 20. Even within this group there were major differences. The annual emission levels of the top five emitters at the time of the study (China, United States, India, Russia, Japan) ranged from 1.3 to 10.5 metric gigatons of CO_2, whereas the emission levels of other group members was still far below the 1-gigaton mark. Second, a significant number of countries had very low, almost insignificant annual CO_2 emission levels. Third, there was a large group of countries with neither extremely high nor negligibly low emission levels. This group deserved its own category rather than being split in the middle and placed with the high or low emitters. Based on these features of the global CO_2 emission distribution, I placed the cutoffs between high and medium levels of emissions at 200 million metric tons of CO_2 per year, and between medium and low levels of emissions at 10 million metric tons per year. Based on these data, 25 countries had high emission levels, 74 had medium emission levels, and the majority—118—had low emission levels. A complete list with cutoffs is included in table A.1 of appendix A.1.

Determining a cutoff between high and low vulnerability levels was equally challenging. Especially in the middle of the distribution of levels, it seemed arbitrary to place countries in different categories. However, given the number of dimensions along which countries' vulnerability profiles differed, a more differentiated approach would have added very little value while imposing major complexity to the group matrix.

Problem 3. The variables I used for the group matrix relied on existing measures of CO_2 emissions and vulnerability, which faced a number of challenges. Data on national greenhouse gas emissions were (and still are) scarce and incomplete. There was (and is) no single data source for all greenhouse gases and all countries.

There were also some fundamental definitional and measurement problems regarding countries' vulnerability to climate change (Adger 2006). The starting point for most attempts to measure vulnerability is the definition provided by the Intergovernmental Panel on Climate Change (IPCC) in its third Assessment Report in 2001: "the degree to which a system is susceptible to, or unable to cope with, adverse effects of climate change, including climate variability and extremes. Vulnerability is a function of the character, magnitude, and rate of climate variation to which a system is exposed, its sensitivity and its adaptive capacity" (IPCC 2001, 995 [Annex B]). One could summarize this definition in a formula: vulnerability = exposure + sensitivity - adaptive capacity. A more recent IPCC definition speaks of "the propensity or predisposition to be adversely affected" (IPCC AR5 WG2, SPM, p. 5). From here, various approaches take off in very different directions, offering diverse vulnerability measures for countries, communities, or individuals.

Measures of exposure tend to be most advanced and are based on geological, physical, biological, or chemical variables for which quantifiable data are available. But, even here, there are major differences between the types of climate impacts social systems can be exposed to (e.g., coastal erosion vs. droughts vs. floods), making comparisons difficult. Further, the methods employed to forecast exposure have significant limitations.

Measuring sensitivity requires the assessment of current resilience of the system, for example, the age and health of a person exposed to a heat wave. This can involve multiple, system-specific, dynamic variables, posing major challenges of quantification and cross-case comparison.

Finally, measures of adaptive capacity—the ability to anticipate, cope with, and adapt to certain impacts—have to consider multiple factors, such as access to knowledge, resource availability, technological advances, and social capital. Adaptive capacity is often equated with level of development. Thus the Least Developed Countries (LDCs) are thought to have the least adaptive capacity. But such an equation might be a misleading simplification. For example, wealthy communities in the United States presumably have high adaptive capacity but might be incapable of—or outright opposed to—addressing climate change–related challenges for various reasons specific to highly developed countries. Thus ideological forces like the Tea Party might rally to defeat adaptation initiatives or to encourage behavior that increases climate change risk exposure, as when North Carolina's legislature ruled out consideration of possible climate change–related sea-level rise by state planners in order to encourage continued development along the North Carolina coastline (Phillips 2012). Less developed communities in poorer countries, on the other hand, might have strong social capital and know how to leverage international support for adaptation, greatly reducing their vulnerability to climate change.

Since vulnerability is a dynamic phenomenon, shaped by the interplay between constantly changing biophysical and social processes in complex systems, it is "not easily reduced to a single metric and is not easily quantifiable" (Adger 2006, 274). Measuring vulnerability involves evaluating a combination of social and environmental data as well as projections of future changes. Generalized measures of vulnerability therefore are often composites of different components (e.g., community type, time period, climate stimulus) that seek to capture certain dimensions of vulnerability, and often have a certain purpose in mind such as disaster risk reduction. The rationale for aggregating various measures is often unclear, which gives rise to significant differences across indexes and rankings. In addition, normative considerations are often interlaced with the analysis. Further, vulnerability measures at the country level cannot capture the differences in vulnerability experienced by communities, organizations, and individuals within that country. Finally, most existing studies, by focusing on developing countries, do not allow for consistent global comparison.

To sum up, the available indexes and rankings for vulnerability to climate change need to be approached carefully, distinguishing the indicators and aggregation methods used, as well as motivating normative perspectives. Since the use of a vulnerability index for this study had a very limited purpose—identifying study participants with different points of view—I canvassed a range of existing vulnerability indexes[1] and chose

1. I canvassed seven indexes in all: the University of Notre Dame Global Adaptation Index (ND-GAIN) by the Global Adaptation Institute; the Global Climate Risk Index by Germanwatch; the Climate and Regional Economics of Development's Vulnerability Index (VI-CRED) by the Stockholm Environment Institute; the

to work with two indexes that combined a general assessment of adaptive capacity (ND-GAIN) with data for 2010 and exposure to extreme weather events and natural disasters (Global Climate Risk Index) with data up to 2011.

The ND-GAIN index for 2010 (then only GAIN, http://index.gain.org/ranking) provided a ranking of 161 countries (cutoff between low and high vulnerability at 80) according to an aggregate measure that combines vulnerability to climate change–related hazards and readiness to adapt to the challenges posed by climate change; higher rankings indicate lower vulnerability. The Global Climate Risk Index (CRI, http://germanwatch.org/en/cri) is focused on the impacts of extreme weather events, using data gathered by the global reinsurance company MunichRe. Rather than forecasting into the future, the index is an assessment of past and current damages caused by floods, droughts, hurricanes, and other extreme weather events. In 2012, it offered two rankings for 183 countries (cutoff at 91): one for average numbers over the previous twenty years (1991–2010), and one for 2010, with higher rankings indicating higher vulnerability.

The results of the selection process for diplomats are presented in table 4.2. The thirty-six participant diplomats, among them twelve delegation heads and eleven women (31 percent), represented thirty different countries. The remainder of the book will refer only to the groups these individuals belonged to (participant groups 1-HH–6-LL); not to their country of origin, or position within the delegation.

Study participants included representatives from member countries of nearly all major UNFCCC negotiation groups at the time of data collection: the Umbrella Group, the European Union, the G77 and China, the Alliance of Small Island States (AOSIS), Small Island Developing States (SIDS), the Least Developing Countries (LDCs), the Bolivarian Alliance (ALBA), the BASIC Group (Brazil, South Africa, India, China), the Environmental Integrity Group (EIG), the Coalition of Rainforest Nations (CRN), and the then newly formed Alliance of Independent Latin American and Caribbean States (AILAC). A number of relevant groups and countries are missing,

Climate Vulnerability Monitor (CVM) by DARA; the Environment Vulnerability Index (EVI) by UNEP/SOPAC; the Structural Vulnerability Assessment (SVA) by FERDI; and the Geographic Distribution of Climate Vulnerability by SEDAC, Columbia University.

Table 4.2
States represented by diplomat participants

		CO$_2$ emissions 2010			
		High[a]	Medium[b]	Low[c]	
Vulnerability	High	Brazil	Bangladesh	Botswana	
		Indonesia	Bolivia	Dominica	
		South Africa	Colombia	Grenada	
			Guatemala	Mozambique	
			Philippines	Namibia	
			Pakistan	Samoa	
				Uganda	
	Subtotals	3	6	7	16
	Low	Australia	Argentina	Barbados	
		Canada (2)	Denmark	Cape Verde	
		Germany	Finland (2)	Iceland	
		Japan	Singapore		
		South Korea (2)	Sweden (2)		
		United States (3)			
Subtotals		10	7	3	20
Totals		13	13	10	36

[a]Greater than 200 million metric tons of CO_2.
[b]Greater than 10 million metric tons of CO_2.
[c]Less than 10 million metric tons of CO_2.

among them India, China, and oil-producing Middle Eastern countries, whose delegations did not respond to my recruitment efforts. Many developing countries have recently formed a new negotiation group: the Like-Minded Developing Countries on Climate Change (LMDC). Table A.2 of appendix A.2 details the membership of study participants in the 2012 UNFCCC negotiation groups. Figure 4.1 shows the match between these negotiation groups and the six study participant groups.

Nonstate Actors Since nonstate actors in the UNFCCC negotiation process do not represent countries, but usually speak on behalf of special interests, the selection criteria for diplomats (national emission and vulnerability levels) were not appropriate. Instead, using the list of participating

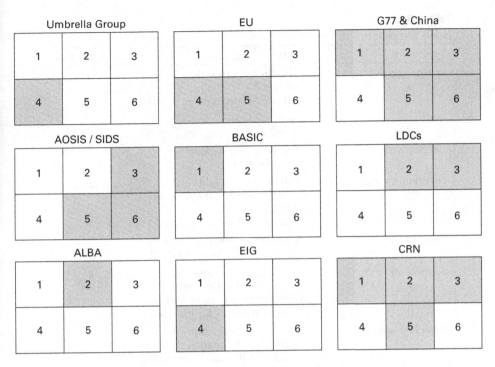

Figure 4.1
Diplomat participant groups and UNFCCC negotiation groups to which they belong.

observer organizations at the COP 17, I developed a simple typology of NGOs and private sector actors based on the interest they sought to protect or advance: youth, faith, development, environment, environment and market, business and technology, fossil fuels, and local or subnational governments. I identified multiple organizations in each category and invited their representatives to participate in the study. Table 4.3 summarizes the results of this process.

The nineteen nonstate actor study participants were of eight different nationalities. Most were from the United States and Canada, but they also included Indian, Brazilian, and South African nationals and eight women (42 percent). Throughout the book, I will discuss the beliefs of these individuals by reference to their nonstate actor type (e.g., youth NGOs).

Based on the alignment between some of the interests of state representatives and nonstate actors, one could expect some overlap between the six-group matrix for diplomats and the eight nonstate actor types identified

Table 4.3

NGOs and private sector actors represented by study participants

Type	Observer organization	Subtotals
Youth	Earth in Brackets (EIB) SustainUS Climate Action Network (CAN)	3
Faith	World Vision Christian Aid	2
Development	ActionAid International Germanwatch	2
Environment and market	National Round Table on the Environment and the Economy (NRTEE) Environmental Defense Fund (EDF)	2
Environment	Amazon Conservation Team (ACT) World Wildlife Fund (WWF) (2)	3
Business and technology	World Business Council on Sustainable Development (WBCSD) Global Carbon Capture and Storage Institute (Global CCS Institute)	2
Fossil fuel industry	World Coal Association (WCA) World Steel Association (WSA) International Council on Mining and Metals (ICMM) Shell	4
Subnational governments	Local Governments for Sustainability (ICLEI)	1
Total		19

above. For the purpose of cognitive-affective mapping, I clustered state and nonstate participants in the following manner (also see figure 4.2):

- Group 1-HH—Youth NGOs;
- Group 2-MH—Development and faith NGOs;
- Group 3-LH—Environment NGOs;
- Group 4-LH—Fossil fuel industry, business and technology;
- Group 5-LM—Environment and market NGOs;
- Group 6-LL—Subnational governments.

The selection process resulted in a group of 55 participants, which included individuals from 32 countries, representing 30 states and 18 different civil society and private sector organizations.

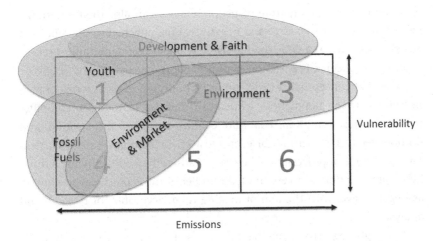

Figure 4.2
Potential alignments of interests and viewpoints between state and nonstate study participants.

Cognitive-Affective Mapping

My analytic approach was grounded in qualitative content analysis. Usually, content analysis relies on text sources, including speech, interview and focus group transcripts, or other publicly available documents. For the purpose of this study, I adjusted the standard approach and placed one analytic step between the collection of text material (interview transcripts) and the application of a code book: I used interview transcripts as a source for the development of cognitive-affective maps, and proceeded to code the CAMs. The cognitive-affective mapping process has a number of important effects on the normal content-analysis process. First, a CAM reduces the information provided verbally by an interviewed subject to key concepts. Transcribed sentences are reduced to single words or phrases. But CAMs also expand the information available to the analyst in three important ways: they provide emotional information, establish structural relations between concepts, and offer visual information that is mentally quite different from the information provided by a linear text source.

Cognitive-affective mapping is a qualitative research tool to identify, visualize, and analyze existing belief structures (Homer-Dixon et al. 2014). A CAM is a network diagram or concept graph that "displays not only the conceptual structure of people's views, but also their emotional nature, showing the positive and negative values attached to concepts and goals" (Thagard 2012, 37). The CAMs' networked representation of sets of connected concepts is based on neural network research in the cognitive sciences that conceptualizes and simulates brain processes in terms of connections between populations of neurons that can be modeled computationally (Galushkin 2007). That said, cognitive-affective maps cannot and do not seek to represent all or even most of the complex biological, neural, and chemical processes of the brain. Rather, they offer a highly simplified representation of a very limited number of those processes at a highly aggregated level with the aim of making them accessible for research and analysis.

Although cognitive maps have been used in the past (Axelrod 1976; Bonham 1993; Novak 1998), there a number of features that distinguish cognitive-affective mapping from previous approaches to cognitive mapping, and make it particularly well suited for this study. The most obvious and theoretically relevant novelty introduced by CAMs is their inclusion of affective information, adding an important layer of data about mental states and processes (Mercer 2010). Affect is the combination of emotion, mood, and motivation. Recent literature in multiple disciplines including psychology, political science, and the decision sciences has emphasized both the need to integrate affect into the analysis of human behavior (Damasio 1995; Loewenstein et al. 2001; Lebow 2005; Vohs, Baumeister, and Loewenstein 2007; Sasley 2011) and the methodological difficulties of doing so (Crawford 2000; Bleiker and Hutchison 2008).

In contrast to Robert Axelrod's mental maps (1976), CAMs focus not just on causal beliefs, but on the network of all relevant concepts for a given subject matter. In the context of global climate change politics, this can include concepts related to the nature of climate change, the meaning of equity, or the values a person considers threatened by climate change impacts. This approach results in a fuller picture of a person's belief system, in other words, the set of ideas and beliefs that serve as motivations or inputs into decision-making processes.

In addition to being more comprehensive, cognitive-affective maps have the unique ability to reveal not only the content but also the structure of a person's belief system. By structure, I refer to the unique constellation of connections between different mental representations, which ultimately creates meaning. These topological features add significant value to the qualitative inquiry into belief systems. Going beyond more conventional text-based analysis, which has to build on a linear ordering of statements or sentences, CAMs "provide an immediate gestalt of the whole system and of the simultaneous interactions between, and relationships among, its parts" (Homer-Dixon et al. 2014, 3–4).

The Nature and Limitations of Cognitive-Affective Maps

The main elements of a CAM are network nodes, which represent discrete cognitive elements, the emotional valences of these nodes, and the links or connectors between any two such nodes, which indicate the relationships between them.

There are four different types of concept nodes, graphically depicted with different shapes and colors. Positive concepts, those with a positive emotional valence ("good" concepts), are depicted as green ovals; neutral concepts, those without any apparent emotional valence, are depicted as yellow rectangles; and negative concepts, those with a negative emotional valence ("bad" concepts), are depicted as red hexagons. Ambivalent concepts, those perceived as positive in some contexts and negative in others, are each depicted with a purple oval and hexagon combined. The thickness of a concept shape's edges represents the emotional intensity associated with the concept node.

Links are lines connecting two nodes, indicating both the emotional and logical relationship between the two concepts, with an emphasis on emotion (see appendix A.3 for details). CAM links are symmetric (undirected), implying that the emotional valences of two connected nodes can mutually influence each other depending on the cognitive process at a particular moment. Which direction dominates at any given moment depends on a number of factors, including priming effects and the emotional strength differential. But because CAMs are static snapshots of a belief system, they do not offer any insights into the cognitive dynamics of emotional valence change or emotional causation. They merely represent the potential of each conceptual node to influence the emotional valence of another.

There are two different types of links in a cognitive-affective map. Solid lines indicate emotional similarity between concepts; they connect two concepts with the same emotional valence (e.g., two positive concepts): if you like one concept, you also like the other; if you do not like one, you also do not like the other. A dashed line represents a relationship of emotional opposition between two concepts, always connecting a positive with a negative concept. Though they depend on the cognitive, often logical, relationship between concept nodes, emotional valences need to be distinguished from that relationship. There is no easy way to categorize the possible cognitive relationship between concepts. It can be based, for example, on constitution, causality, or co-occurrence. The absence of a link indicates the lack of a relationship between the nodes.

Links can have varying strengths. The factors that influence the strength of a link have so far not been explored empirically, but they could include availability and quality of evidence; whether the linked beliefs rest on personal experience or abstract information; uncertainty or newness of a concept; or the source of information about a concept, such as a trusted person with shared values, an expert, or a stranger (Kahan, Jenkins-Smith, and Braman 2011).

All possible node types and some of their possible relationships are displayed in figure 4.3.

The CAMs' way of representing cognitive entities and their connections is strongly influenced by recent cognitive theories, in particular, the theory

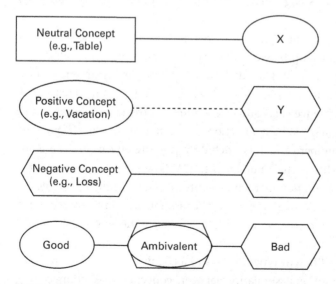

Figure 4.3
Basic CAM elements and link types.

of emotional-cognitive rather than just cognitive coherence (Thagard 2006). Coherence is a macro-level quality of a belief system, an emergent property based on many cognitive-emotional interactions between multiple nodes at the micro level. By focusing on emotional coherence, CAMs are thus highly simplified representations of one aspect of cognition—the emotional structure of belief and value systems—necessarily ignoring many other aspects. This simplification is useful for some analytic purposes but limits what CAMs can and cannot deliver. These and other issues regarding the nature and limitations of cognitive-affective maps are discussed in greater detail in appendix A.3, which contains the standard protocol for addressing them in the context of this research.

Data Source: Semistructured Qualitative Interviews   Cognitive-affective maps can be generated in a number of different ways, including use of software tools such as Empathica (Thagard 2010a).[2] Research subjects can develop their own CAM of a specific issue after an introduction to the CAM technique and the formulation of the topic to be mapped. Alternatively, the researcher can generate the CAM based on data gathered with qualitative interviews, from primary and secondary text sources (e.g., published statements, speeches, journal articles, statutes) or from observations. The validity of such a researcher-generated CAM can be verified in a (second) interview with the research subject, providing the opportunity to correct the map by adding, deleting or changing concepts, emotional valences, or links. Another way to validate an initial text-based CAM is to invite other researchers to generate a CAM based on the same source material, and then compare the results. This subsection and the next outline the approach adopted for this study.

I conducted fifty-five semistructured interviews (see appendix A.4 for the interview protocol) to generate input data (transcripts) for the CAMs. The interview questions covered eight different themes: (1) the science of climate change and expected climate change impacts; (2) the political nature of the climate change problem, its solvability and ideal solution, and obstacles to that solution; (4) interest definition; (5) cost-benefit analysis;

2. Empathica, which is available for free, was created in 2011 by fourth-year software engineering students at the University of Waterloo, Ontario. Learn more at http://cogsci.uwaterloo.ca/empathica.html.

(6) assessment of COP 17; (7) concern about climate tipping points; and (8) imagining a 2080 worst-case scenario of climate change.

The questions were open ended and designed with an important trade-off in mind. On the one hand, it was important to allow study participants to steer the conversation in a direction that was most meaningful to them and to insert as much of their subjective views as possible into their responses. On the other hand, the interview questions provided thematic anchors that generated conceptual clusters open to comparative analysis.

There are a number of insurmountable problems associated with the interview approach to studying individual cognition. First, it is impossible to know or verify whether interviewed subjects are honest in speaking about their personal beliefs. Subjects might present the beliefs of the organizations they represent as their own, or offer opinions they believe are appropriate or expected rather than their own. This is a general problem for social scientific research and not unique to this study.

Finally, the stability of subjects' beliefs over time is unclear, raising the question whether their responses would have been different on any other day or under different circumstances. Various factors influence the interview situation (e.g., mood, stress, an earlier meeting), and can make certain concepts more or less important than they would have been at a different point in time. Since the study interviews were conducted under very different conditions, for example, some in participants' offices, some during the Bonn Climate Change Conference in May 2012, some in person, and some via Skype or telephone, the influences under these different conditions are not comparable, nor even knowable. Some elements of the research design alleviated this problem, however, giving participants the opportunity to present and correct their opinions outside of the interview situation and on their own time (see below). Further, it was possible to cross-reference the interview data and the available Q sort data (see "The Process of Doing Q" below) to spot inconsistencies. Each interview and its Q sort usually took place several weeks or even months apart, and therefore could offer some clues regarding a study participant's belief stability over time. That said, neither the CAM nor the Q study could address the possibility that participants changed their minds during the study period.

Process: Generating Cognitive-Affective Maps   I generated fifty-five CAMs based on interview transcripts, selecting key concepts—single words or short phrases—while assessing their emotional valences based on the context and meaning provided by the interviewed study participants. I used the software Empathica to create network graphs. Words that participants had used in the same sentence or the same paragraph tended to be linked and located close to one another in the map. My interview questions provided seven conceptual anchors (hubs)—"Climate change," "Me," "Science," "Impacts," "Problem nature," "Solution," "UNFCCC"—which I used to create a similar spatial pattern across all CAMs. Figures 4.4 to 4.6 illustrate this process, building a CAM in three steps.

I placed the concept "Climate change" at the center of each CAM generated for this project. Concepts such as "Science," "Impacts," "UNFCCC," or "Solution" were the focus of various interview questions and served as conceptual hubs for building the CAMs. They all have a direct link to "Climate change." "Climate change" is a negative concept (depicted as a hexagon), whereas the concept "Solution" has a positive emotional valence (indicated by an oval). Initially, most concepts that had no clear emotional valence, such as "Science," were depicted as neutral (rectangles). Their emotional valence could be adjusted later, depending on the emotional valences and influences of other concepts it became linked to in the process of developing

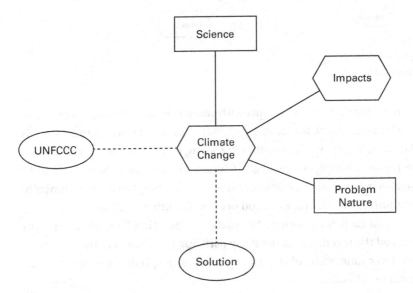

Figure 4.4
Sample CAM, step 1.

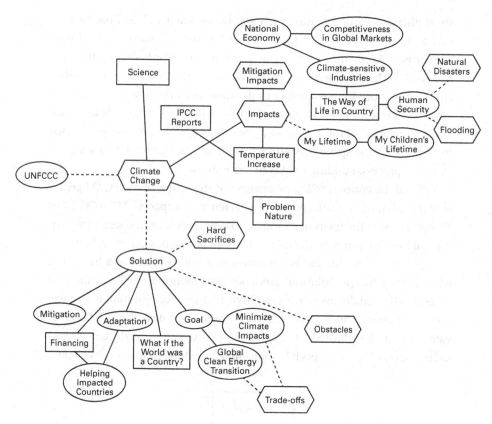

Figure 4.5
Sample CAM, step 2.

the more detailed map. Concepts with similar emotional valences (e.g., the negative concepts "Climate change" and "Impacts") were connected with solid lines, implying a mutually reinforcing emotional effect. The dashed line between two concepts (e.g., "Climate change" and "Solution") indicated their emotional tension; often these opposing forces were mutually constitutive, that is, one was good because the other was bad.

Around each hub, such as "Impacts" or "Solution," multiple concepts indicated the specific ideas the study participant discussed during the interview. Over time, some of the concepts linked to specific hubs also became connected globally.

In order to verify that the CAMs generated this way reflected the thoughts of the research participants well, I sent all participants their draft CAMs,

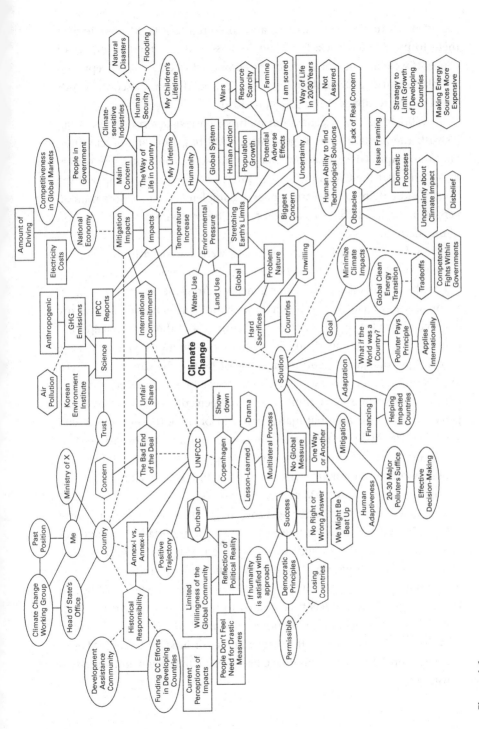

Figure 4.6
Sample CAM, step 3.

provided instructions for reading and interpreting a CAM, and invited them to provide feedback. For example, I asked them whether important concepts were missing, or whether the emotional valences of concepts and the placement and strength of links were correct. I then revised the CAMs based on the feedback participants provided.

To validate my CAM generation method, I asked two colleagues familiar with the method to each generate a CAM based on one of the interview transcripts. Given that the interview transcript contained about 6,000 words and that the CAMs I generated rarely exceeded 150 concepts of the usually more than 1,000 possible ones, a complete match of more than 50 percent of selected concepts could be considered a satisfactory validation of the method.

The CAMs produced by my colleagues were significantly less detailed than my own. Indeed, with 68 and 62 concepts, they were less than half the size of the original CAM that I had generated from the selected transcript. I limited the comparison to the total number of concepts contained in my colleagues' two CAMs, 95. My CAM contained 74 of these 95 concepts (78 percent). A similar comparison was made for the emotional valences of the 77 concepts that were identified by at least two of the three researchers. The emotional valences assigned by the researchers completely matched (all three researchers agreed) in 46 cases (60 percent). For 18 concepts (23 percent), there was a partial match (two out of three researchers agreed), and there was disagreement regarding the remaining 13 concepts (17 percent). Although identifying emotional valences appeared to be more challenging than identifying relevant concepts, both tests offered a sufficiently strong validation of my CAM generation method.

Nevertheless, when it came to generating CAMs with high empirical validity, I faced a number of challenges. The most important challenge was lack of data about the strengths of links in a CAM. Having no data that would allow me to quantify the strength of the links, my setting of weights was strongly driven by my intuition and background knowledge. Although I had requested specific feedback related to the weight of particular links from my study participants, I did not receive any. This came as no surprise. Providing such feedback for the large number of links in their CAMs (on average, more than 100) would have been both challenging and very time consuming.

I faced similar challenges with regard to the strength of emotional valences of the concept nodes. In some cases, the participants' emotive language (e.g., "horrible," "extremely frustrating," "fear") helped distinguish stronger affective valences from weaker ones. But time constraints and the type of study participants (see below) made it impossible to gather sufficient data to assign all emotional valences based on satisfactory empirical evidence.

All anonymized fifty-five CAMs are available at https://mitpress.mit.edu/books/mindmade-politics/.

CAM Content Analysis   As in the standard approach to qualitative content analysis, I used a coding scheme to identify and categorize existing concepts, meanings, and emotional information in the cognitive-affective maps. The aim of the coding effort was threefold: to identify the most relevant concepts in existing belief systems, to assess the extent to which cognitive assumptions of major international relations theories matched the cognitive reality in the case of global climate change politics, and to understand the role (if any) of the unique characteristics of climate change in the belief systems of individuals involved in the global political process.

The coding scheme (see table A.3 of appendix A.5) had three parts. The first sought to identify concepts related to theoretical expectations of the decision-making models of major international relations theories (e.g., costs, benefits, identity, norms). The second part focused on the special characteristics of the climate change problem. This included concepts related to complexity, the existence of climate tipping points, long-term climate change impacts, and responsibilities to future generations. Finally, the coding effort sought to identify other concepts that might have been relevant but that did not fit the two categories above (e.g., agency, hope). The analysis proceeded in clusters, using the six participant groups as a clustering device that enabled in-group as well as cross-group comparisons (discussed in greater detail in the following section).

Q Method

Created by William Stephenson (1935) for the systematic study of subjectivity, the Q method is a well-established research technique. A Q study enables the researcher to identify and explore the given viewpoints on a specific topic within a certain population of subjects, by analyzing patterns (underlying "factors") in the opinions of subjects. These opinions are elicited in a special type of survey that requires subjects to rank-order a given set of statements about the topic. The statistical analysis of these rankings reveals so-called factors—viewpoints—that are shared by several individuals. In my study, I call these factors "belief systems"—relatively stable, linked, and emotionally coherent sets of ideas related to a specific subject matter.

Applied in multiple fields, including psychology, political and policy sciences, health and environmental studies, the Q method uncovers a high degree of qualitative detail about individual and collective thinking and allows the researcher to gain a holistic understanding of existing belief systems—the substantive gestalt of a point of view rather than a structural one. At the same time, it is driven by statistics (inverted factor analysis), offering a methodological bridge between qualitative and quantitative approaches. The quantitative basis of the analysis differentiates the method from other, more textual approaches to exploring beliefs, but also differs from other quantitative techniques such as opinion polls and surveys.

A Q study elicits different viewpoints regarding a specific subject matter; in the case of my study, multilateral cooperation on climate change. The Q method does not require any advance knowledge or hypothesis about existing belief systems, which simply emerge from the data. Although the researcher does not know the number of existing viewpoints at the outset, the Q method assumes that, regardless of the issue domain, the number of possible viewpoints is finite, generally smaller than seven (Barry and Proops 1999, 339), and normally about five (Brown 1980, 62). But Q studies with only two or three factors are also common.

In my study, I developed a solution of six belief systems. This solution was entirely independent of the division of participants into six groups, which was necessary for participant selection and the CAM analysis. The fact that there were six participant groups and six belief systems was a coincidence or, more correctly, a result of the data structure. The groups and belief systems did not overlap, in other words, it was not the case that all members of a participant group (as defined by me for the purpose of recruitment and analysis) shared the same belief system or systems.

In a Q study, subjects attribute their own personal meanings to a fixed set of statements (stimulus items), which they rank on a prearranged grid that approximates a normal distribution, with few statements toward the extreme ends of the scale and most statements in the center. This sorting exercise reveals subjects' individual beliefs and belief structure through their interaction with the statements, their operant subjectivity. By correlating individuals rather than objective traits in an inverted factor analysis (Watts and Stenner 2012, 7–14), the correlation of Q sorts provides information about similarities and differences in the structured beliefs of groups of people.

In my Q study, the emergent groups of participants, that is, those sharing a viewpoint, differed from the groups I predefined for the purpose of participant selection, and also from the existing UNFCCC negotiation groups. This observation might open up interesting opportunities for changing political dynamics based on existing beliefs rather than national interests or positions of negotiation groups.

More broadly, the Q method permits a conception of subjectivity that is complex and multifaceted in the sense that individuals may not only hold more than one clearly bounded perspective; they may also share elements of a number of different perspectives (Harré and Gillett 1994, 25; Hajer 1996). Binary categories (e.g., male vs. female, or developed vs. developing) are often unable to capture this complexity. The Q method provides opportunities to explore these multidimensional, overlapping perspectives.

Critiques of Q studies often attack the method for not producing results representative of the larger population, which, however, it does not claim it can do in the first place. The limited aim of a Q study is to identify existing viewpoints based on individuals' internal frames of reference within a given community. It does not and cannot make any claims about the characteristics of the population at large (e.g., all those engaged in global climate change negotiations), such as the percentage of the larger population associated with different factors or viewpoints. A Q study does not claim to be representative in the sense that all possible viewpoints have been identified, or even that its results are indicative of the importance and distribution of viewpoints within the community under consideration (Dryzek and Berejikian 1993, 51). But the method has proven its ability to quickly unearth the most relevant perspectives within such a community on any given issue.

Conventional Q studies identify beliefs without paying specific attention to the associated emotions. Since the Q method is flexible in this respect, however, I adjusted the research design of my Q study to include an emotional dimension in the analysis. I added statements about emotional states associated with concepts and beliefs and placed special emphasis of affect in the factor interpretation stage.

The Process of Doing Q

A Q study proceeds in five steps. First, the researcher selects the research topic and the relevant community of those individuals whose viewpoints are of interest and require exploration. Second, using a variety of sources, the researcher identifies the full range of opinions on the topic that are relevant to the selected community. From the universe of possible statements, often called the concourse, the researcher selects a Q set—a limited number of representative statements that capture the largest possible opinion space and that maximize the ability of all research subjects to express their views. This is a deliberate effort to create boundaries around the research topic. Third, facilitated either by printed statement cards or by an online platform, each subject is asked to perform a Q sort—to respond to the Q set by rank-ordering the statements on a predetermined scale based on the intensity of the subject's own agreement or disagreement with these items. Fourth, statistical analysis of these Q sorts enables the researcher to extract factors—patterns in the data based on the similarity of statement rankings that identify different points of view. The initial factor extraction is complemented by factor rotation—the maximization of differences between factors that help explore various hypotheses. For example, factor rotation can be used to single out individuals who have a special role within the investigated community. The Q sorts of individuals who have strong factor loadings are used to calculate the factor scores—standardized Q sorts representing specific factors. Finally, the resulting factors require interpretation. The systematic analysis of the content of these different viewpoints, their most important differences, and the individuals associated with a certain viewpoint can create a deep, holistic understanding of the research topic explored.

Application

I tested the Q study design in a pilot study with forty-four participants from ten countries, none involved in the global climate change negotiations.

Based on the pilot results, I adjusted the number and content of items in the Q set.

Concourse, Q Set, and Grid For the creation of a concourse—a set of statements that reflect all possible opinions on the subject matter within the relevant community—I relied on a range of text sources, including the coverage of the climate change issue in the print media (US, Canadian, and German newspapers), online sources (UNFCCC website, blogs, NGO websites), journal articles, and discussions with colleagues researching climate change politics. These were honed down to the final Q set in a process of analysis (e.g., of similarities), testing (the pilot study),[3] and workshops. The final list of statements contained sixty-five items in eight different categories: (1) science and causes; (2) problem nature; (3) actors; (4) policy goals and options; (5) economics and development; (6) identity; (7) ethics and justice; and (8) special characteristics of the climate change problem (see appendix A.6). Based on the number of statements, I chose the grid structure with a score range from –5 to +5 (see figure 4.7).

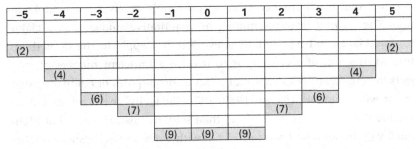

Figure 4.7
Frequency distribution for Q sort.

3. Using the results of the pilot study, I identified a number of statements that yielded very little additional insight, for instance, either because all study participants ranked them similarly (no differentiation), because the statements captured important views of a domestic rather than global discourse, or because no study participant considered them important. These were removed from final the Q set. I changed and added some statements based on the written responses of pilot study participants.

P Set Two main goals drove my selection of participants from the community of interest, the P set: capturing a minimum diversity of existing viewpoints, on the one hand, and including viewpoints that mattered in relation to the research issue, on the other (Watts and Stenner 2012, 70–71). Because strategic rather than random sampling is desirable in forming a P set, I made use of an initial hypothesis and existing knowledge about the issue, while taking into account practical constraints specific to the sample group.

In selecting participants for this project, I emphasized diversity of viewpoints over importance. I invited all fifty-five individuals who participated in the interviews for the CAM analysis (see "Participant Selection" above) to take part in the Q study; twenty-eight individuals accepted the invitation, which is not an unusual P set size in published Q studies. In order to control for and reveal my own bias when studying belief systems, I participated in the Q sort myself, which resulted in a total P set of twenty-nine.

Given the complexity of the issue at hand and the corresponding potential for a diversity of views, how I arrived at a number of study participants that was both necessary and desirable deserves some explanation. The logic of large-n sampling in quantitative studies with a claim to universality and representativeness does not apply to the Q method. Instead, the aim of a Q study is to involve a sufficient number of subjects to "establish the existence of a factor for purposes of comparing one factor with another" (Brown 1980, 192). Generally, the P set should be smaller than the Q set (here, fewer than sixty-five participants), but high-quality Q studies have been carried out with as few as twelve subjects (Niemeyer 2004). Given the complexity of the climate change issue, I assumed that there might be a relatively high number of viewpoints (from five to seven), possibly closely related to the six groups defined for participant selection (see "Participant Selection" above). Since it would be desirable to have at least two individuals with similar perspectives and consequently similar Q sorts, from ten to fourteen participants might have been mathematically sufficient, but every additional participant increased the probability of including a broad range of viewpoints and of having at least two participants share a viewpoint.

There are two reasons why some uncertainty regarding the quality of the statistical analysis remains. First, participants' viewpoints could not be known before selection, especially given the possible difference between the participants' publicly presented negotiation positions and privately held beliefs. Second, participants self-selected into the Q study. For example, it was possible that half of the P set shared the same viewpoint, increasing the chances that some viewpoints captured in the CAMs would not be represented at all, or that a single Q sort would become a factor. However, these concerns did not prove to significant because the final P set was a fairly good representation of the larger set of study participants, containing fifteen diplomats, with at least one individual from five out of the six participant groups (group 1-HH is missing), and thirteen NGO participants from seven out of eight different NGO types (those from subnational governments are missing).

Sorting Q sorting was conducted online. In addition to the ranking exercise itself, the Q survey asked participants to provide basic demographic information and to comment on the statements they had ranked as +5 or –5. It also asked them whether they could not find particular statements to express their respective points of view, and to identify which statements had been particularly difficult to place in the grid (e.g., because of conflicting interpretations). Appendix A.7 contains the instructions for the Q sorting exercise and the list of additional questions.

Factor Extraction and Rotation For the statistical analysis, I used AdvanceQ, software provided free of charge by Simon Niemeyer and David Moten (2008). I developed the factor solution in four steps: (1) extraction of six factors using the Centroid method; (2) factor rotation (Varimax with manual adjustments); (3) identification of Q sorts with significant loadings (0.36) on each factor;[4] and (4) calculation of factor scores based on these significant loadings.

Deciding how many factors to extract in a Q study can be guided by many criteria. The standard approach is to extract only those factors which

4. Factor loadings are correlation coefficients indicating the degree to which a sort correlated with a factor. The significance threshold at a 99 percent confidence level was calculated by multiplying the standard error (SE) by 2.58, where $SE = 1/SQRT$ and $SQRT$ = the number of statements in the Q set (65). Using these calculations, the significance level for this study was 0.397. I lowered the significance level to 0.36, allowing for the inclusion of the sorts of three participants (13, 15, 21).

have an Eigenvalue (sum of all squared loadings of a factor) greater than 1.0, which have at least two significant factor loadings, or both (Watts and Stenner 2005, 105). The factor solution I used fulfills both of these criteria, but also emphasizes simplicity, in other words, it minimizes the number of factors without losing a significant amount of information.[5]

Table 4.4 presents the resulting factor loadings, Eigenvalues, variance explained, and communality (percentage of a Q sort associated with the other sorts, h^2). The loadings of Q sorts associated with a factor are marked either by shading or framing; framing indicates confounded sorts, those with three significant factor loadings, (which were excluded from the calculation of factor scores).

The process of factor extraction and rotation revealed that the data set contained a large number of confounded sorts, which was unusual compared to other Q studies. The result was consistent across several solutions, including principal component analysis (PCA) and Centroid factor extraction, a five-, six-, seven-, or eight-factor solution, and multiple rotation options. The solution developed through multiple rotations contained seven sorts that loaded on a single factor only. Most sorts (fifteen) had two significant factor loadings; six sorts showed significant correlations with three different factors.

In the Q method literature, the treatment of confounded sorts is almost nonexistent. The standard approach relies solely on pure sorts that load only on one factor, excluding all confounded sorts because they "muddy the waters" since they offer no clear factor association (Stephenson 1953, 107–109; Brown and Robyn 2004, 114–117; Akhtar-Danesh, Baumann, and Cordingley 2008). Although a detailed discussion of confounded sorts is lacking in the literature, there are strong arguments for including them in the creation and interpretation of factors, or at least for giving greater theoretical attention to the issue of including them. The existence of confounded sorts merely reflects the reality of human thinking. People often hold or share parts of different points of view, and each point of view can

5. I tested eight-, seven-, six- and five-factor solutions. Even when I extracted eight factors, some solutions would meet the Eigenvalue and significant loadings criteria. However, the similarities between some of these eight factors were too great to offer interesting insights; presenting minor variants of the same factor. The six-factor solution offered the best compromise in terms of maximum differentiation and minimum redundancy.

Table 4.4
Rotated factor loadings

Participant ID	Factor loadings						h^2
	A	B	C	D	E	F	
1	0.23	0.21	0.27	0.18	0.42	0.36	0.51
2	0.19	0.20	0.28	0.59	0.48	0.20	0.78
3	0.33	0.59	0.09	0.25	0.37	0.09	0.67
4	0.31	0.43	0.01	0.25	0.27	0.47	0.64
5	0.01	0.27	0.09	0.47	0.29	0.48	0.61
6	0.52	0.28	0.45	−0.01	0.50	0.21	0.84
7	0.33	0.08	0.05	−0.26	0.00	−0.09	0.19
8	0.11	0.50	0.22	0.05	0.56	0.19	0.67
9	0.17	0.17	0.59	0.13	0.20	0.18	0.50
10	0.54	0.24	0.26	0.27	0.51	0.08	0.76
11	0.33	0.27	0.62	0.22	0.20	0.26	0.72
12	0.51	0.41	0.19	0.45	0.30	−0.02	0.75
13	0.28	0.39	0.36	0.32	0.22	0.22	0.55
14	0.14	0.55	0.40	0.40	0.30	0.03	0.74
15	0.25	0.36	0.13	0.26	0.38	0.26	0.49
16	0.70	0.25	0.23	0.41	0.12	−0.04	0.79
17	0.28	0.43	0.61	0.26	0.09	−0.04	0.72
18	0.43	0.04	−0.01	0.15	0.36	0.23	0.39
19	0.41	0.04	0.32	−0.15	0.11	0.20	0.35
20	0.37	0.15	0.41	0.45	0.20	0.32	0.67
21	0.32	0.20	0.31	0.37	0.18	0.27	0.48
22	0.40	0.43	0.31	0.07	0.11	0.44	0.64
23	0.13	0.17	0.33	0.48	0.58	0.01	0.73
24	0.14	0.73	0.11	−0.01	0.08	0.10	0.58
25	0.52	0.10	0.42	0.52	0.24	0.07	0.79
26	0.41	0.23	0.24	0.43	0.15	0.31	0.59
27	0.28	0.10	0.66	0.05	0.25	−0.11	0.61
28	0.29	0.43	0.31	0.55	0.20	0.26	0.77
29	0.17	0.43	0.31	0.19	0.08	0.24	0.42
Eigenvalue:	3.5	3.5	3.4	3.1	2.8	1.7	
Variance:	12.2	11.9	11.9	10.8	9.5	5.7	
Total variance explained (%)						62.0	

get activated by different circumstances. Examples of individuals with confounded sorts are conflict mediators, who can associate with two or more opposing parties, and ideological moderates, who share parts of a conservative and a liberal perspective. Depending on the Q set, different beliefs get activated in the course of Q sorting and become part of the data.

My Q study data set suggests that relying on pure sorts only would have provided a severely limited picture of the study participants' points of view by removing more data from the analysis than it included. There could be several reasons why confounded sorts were a dominant feature of the data. First, it might reflect my study's particularly complex research issue—climate change—where multiple overlapping beliefs were possible and occurred with a greater likelihood. Second, confounded sorts might indicate that the climate change issue was not yet settled and that strong and clearly distinguishable camps had not yet formed. Finally, it might reflect that, rather than representing a comprehensive point of view, a factor represented only one of several parts of a viewpoint. If so, multiple factors could be part of a person's viewpoint and several factors might be needed to understand that person's larger belief system. These parts of a viewpoint might be combinable in several ways, leading to different patterns of confounded sorts.

Whatever the underlying reason for the large number of confounded sorts, they might offer bridges between possible belief systems and current negotiation positions—avenues along which cognitive change could be fostered. For example, if person 1's sort loads on factor A and B, and person 2's sort loads on factor A only, person 2 might find it comparatively easy to take on some of the beliefs of factor B.

A factor score is created by calculating the weighted average score for each statement from all sorts that load significantly on the factor. My study's factor scores are included in appendix A.8. Due to the nature of this specific data set, I further constrained the calculation of factor scores by excluding all sorts with more than two significant loadings on different factors, although this exclusion of confounded sorts did not affect their impact on factor extraction.

Five sorts, each of which had three significant factor loadings, were excluded—those of one NGO representative of the environment and market type and another of the business type, those of two diplomats in group 4-HL, and those of one diplomat in group 2-MH. The exclusion at the last stage of the mathematical analysis also has implications for interpreting the resulting factor scores. In addition to missing the perspectives of a number of individuals who took part in the CAM analysis but not in the Q sort (e.g., diplomats from countries in the BASIC Group, AOSIS, and ALBA and most LDC diplomats), this exclusion limited the breadth of views represented in the resulting factor solution. Consequently, the factors identified with the Q analysis were mainly expressions of perspectives in participant groups 2-MH, 3-LH, 4-HL, 5-ML, and among youth NGO representatives.

Interpretation   Following Simon Watts and Paul Stenner's approach (2012) to factor interpretation, I used four key categories to compare different factors with one another: (1) items ranked +5; (2) items ranked higher in factor array X than in any other factor array or equally high; (3) items ranked lower in factor array X than in any other factor array or equally low; and (4) items ranked −5. Items in the middle of the distribution deserved special attention because they could have different meanings. Usually Q studies assume that an item ranked 0 has no meaning for the factor because the subjects neither agreed nor disagreed with the statement in this item and were therefore indifferent regarding its content. However, different interpretations of rankings in the center of the grid are possible. Some participants in my Q study may have felt deeply conflicted regarding a particular statement because they had two different interpretations of it, each eliciting a different level of agreement or disagreement. This would be a case of too much rather than no meaning. Further, it is possible that participants agreed with a large number of items and found themselves forced by the grid to place some of these items in the center or even to left of the 0 column.

Synergies
Cognitive-affective mapping and the Q method each have important limitations, but their joint application remedies many of their individual shortcomings. CAM zooms in on subjects' cognitive states and allows them to present their beliefs in detail, using their own language, concepts, and arguments. It maximizes idiosyncratic subject input at the cost of limiting comparability of findings across subjects. The Q method's focus, on the other hand, is on cognitive features that are shared by several subjects.

The method sacrifices detailed information about the individual in order to detect collective points of view.

The primary advantage of combining the two methods is the generation of complementary insights, similar to taking both a micro and macro perspective of subjectivity. Using both tools in conjunction also allowed me to synthesize their respective insights. Observing the reduction of conceptual diversity contained in the CAMs during the scale transition from the individual to the group level, I was able to separate "popular" cognitive elements from those which were primarily expressions of individual preferences and personalities. Further, uncovering the diversity of beliefs in the form of participant CAMs associated with a Q factor provided some understanding of the level of ideational plasticity of a shared perspective.

Finally, each method worked with different constellations of groups of participants; together, they provided insights into the demographic features associated with certain belief systems. The CAM analysis required the design of participant groups prior to the analysis, even prior to participant recruitment. The Q method, on the other hand, allowed groups to emerge based on their shared beliefs about the research issue. The comparison of these groups offered some early insights regarding the nature of existing belief systems, the possibility that parts of belief systems were shared with a number of different groups, and the possibilities for cognitive change. All of this will become clearer in chapters 5 and 6.

5 Four Cognitive-Affective Lessons for Global Climate Change Politics

Cognitive-affective maps can provide a number of different insights into an actor's mind: the content or substance of beliefs regarding a certain issue (i.e., discrete cognitive elements); the type and intensity of emotions associated with these beliefs; the structure of the belief system (i.e., the relationships between discrete cognitive elements); and what one might call the narrative that provides coherence to this belief system.

This chapter lays out the multiple insights derived from analyzing study participants' cognitive-affective maps. These maps were generated in 2012—just after the Ad Hoc Working Group on the Durban Platform for Enhanced Action (ADP) launched a new negotiation process (Decision 1/CP.17) that would result in the Paris Agreement in 2015. Reflecting on the importance of that early phase of the ADP's work and the search for novel ideas and frameworks to overcome the deep divisions between negotiating blocs, this chapter offers a post-Paris analysis of the cognitive landscape of 2012. Which ideas survived and now shape the Paris Agreement? How did these ideas change between 2012 and 2016? Which concerns were addressed, which interests were ignored, which hopes were dashed?

The comparative analysis of all study participants' CAMs revealed four distinct cognitive-emotional patterns that have shaped the reality of climate change negotiations as much after Paris as they had before. The most important pattern concerns both the structure and the content of study participants' belief systems, which were organized around a cognitive triangle—a patterned relationship between three sets of cognitive elements. Then there were two content-specific patterns: one concerning the structure of the international system; the other relating to climate science and complexity, what might be called knowledge. Finally, a fourth pattern revealed by the participants' CAMs concerned the relationship between their thoughts about agency and the emotional phenomenon of hope.

I focused my analysis on questions raised by international relations scholarship, in particular, whether the rational-choice assumption matched the cognitive reality of those engaged in climate change negotiations. But I was also interested in how study participants grappled with the special characteristics of climate change and climate science. Jointly, these insights enrich our understanding of agency in global climate governance.

By the end of the chapter, I hope to have convinced you that the major schools of thinking and theorizing in international relations—neorealism, neoliberal institutionalism, and social constructivism—though often perceived as rivals, in fact complement one another at the cognitive level. In particular, the theoretical interactions between rational-choice and constructivist approaches to international relations offer powerful insights into political decision making in that realm. Jointly, they generate political actors' threat assessments and formulations of their national interests. I also point to the need to expand existing theories of agency to account for time (memory and anticipation) and hope in political decision making and behavior.

The Cognitive Triangle: Threat, Identity, and Justice

Much scholarship on international politics is based on the assumption that political actors are rational.[1] The rational-choice framework assumes that actors can and do calculate the expected costs and benefits of their available action options in a certain situation and choose the option having the greatest net utility, often understood in terms of power or national interests. Social constructivists have criticized this highly simplified and often not very realistic framework for analyzing international politics. Instead of quantifiable costs and benefits, constructivists emphasize norms, identities, and, more generally, ideas as explanatory factors in political decision making. These two theoretical approaches are often considered mutually exclusive: either decision makers calculate net benefits or they are guided by behavioral norms rooted in their identities. And the proponents of

1. Parts of this section were previously published in Milkoreit 2014.

each approach can easily point to cases the other approach simply fails to explain.

Reality is much more complex than either camp suggests. Neither rational-choice theory nor social constructivism is wrong, but, at the same time, neither captures the full picture. In my proposed cognitive-emotional model, rational cost considerations, normative reasoning, and identity constructs interact with one another. Their interaction is facilitated by actors' identity conceptions—especially group loyalties—these three cognitive elements together enable political reasoning and choice.

Moving away from "either or" explanations, I argue that those engaged in global climate change negotiations display patterned interactions between rational thinking and other cognitive elements and processes. These other cognitive elements include mental representations that form the identity of a person or group and that involve moral reasoning. Most likely, similar patterns exist in the minds of political actors outside the sphere of climate change negotiations, perhaps even in the mind of any person who has to make decisions of a certain nature.

I call this pattern a cognitive triangle, linking three clusters of interacting cognitive elements (see figure 5.1). I label the three corners of the triangle threat, identity, and justice. All major explanatory tools developed in the history of international relations scholarship inform the triangle; each offers insights concerning at least one of its component clusters. But to produce a serviceable cognitive-emotional model of political choice, we need to identify and study the interaction patterns between rationality, identities, and justice, in other words, we need to define the dynamic interaction rules that determine cognitive outcomes, even though these interaction rules are less well understood than the elements they operate on.

Based on the data gathered for this project, global climate change negotiators display one of at least two distinct cognitive patterns. In other words, the cognitive triangle has two different settings, which could be called hot concern and cold calculation (Milkoreit 2014). Below I explore each corner of the triangle in greater depth before discussing the hot and cold patterns of thought it generates.

Threat Perceptions

In response to specific interview questions, study participants expressed their concerns about a large and diverse set of threats related to climate

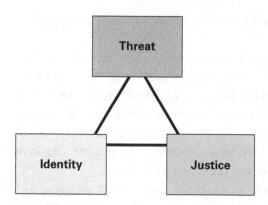

Figure 5.1
Cognitive triangle.

change—from existential threats, identity loss, and human suffering to economic and financial threats related to climate change policy, affecting levels of income and unemployment or national economic performance. Before delving into the different types of threats, it is worth clarifying the difference between threats, risks, and costs as I use these terms in the context of climate change.

The subject of a well-established field of scholarship, risks are defined as threats to human values that have a calculable probability and equally calculable impacts (costs or damages) associated with them (Kaplan and Garrick 1981). Although the risk framework has become increasingly popular in the climate change discourse over the last few years, climate change presents a range of threats whose probability and costs cannot be readily calculated. Given this partial lack of quantifiability, it is more appropriate to speak about the more general concept of climate change–related threats.

A key concept in the rational-choice framework, costs express the negative effects of certain events in monetary terms and provide the foundation for calculating the expected net benefits of different action options. Although some of the negative impacts or losses related to climate change can be readily expressed in terms of dollars or euros, there are important types whose costs are hard, if not impossible, to quantify. These include the loss of life, identity, or culture, the disappearance of islands or the extinction of species.

Currently, the difficulties of categorizing the various threats emanating from climate change are mirrored in UNFCCC negotiations about loss and damage, a relatively novel item on the convention agenda, although small-island states have been fighting to put it there ever since the UNFCCC went into force in 1994. The term "loss and damage" attempts to capture climate change impacts that cannot be prevented by mitigation and that exceed the adaptive capacity of the affected community. The 2013 Conference of the Parties (COP 19) in Warsaw established the Warsaw International Mechanism for Loss and Damage (WIM), which was strengthened by Article 8 of the Paris Agreement. One of the WIM's major areas of work has been to define different categories of loss and damage in order to increase understanding of the phenomenon and enable discussion of appropriate responses. These categories include noneconomic losses and damages (NELDs), which are highly diverse and context specific but, again, hard to quantify (Serdeczny, Waters, and Chan 2016), and thus also hard to integrate into conventional policy and decision making, which rely heavily on cost-benefit analysis. Although the idea of NELDs had not come to light when I conducted my study interviews in 2012, losses of this kind feature prominently in my analysis below.

Given these definitional challenges, I use the term "threats" to denote the broad range of possible negative effects of climate change and of policy responses to climate change, including both economic costs and financial losses and less tangible negative effects, like the loss of culture or human suffering.

Apart from a fundamental distinction between threats related to climate change impacts—the costs of inaction—and those related to action on climate change, it is possible to devise a hierarchical taxonomy of threat categories based on the various descriptions study participants offered during their interviews. Table 5.1 lists seven categories in descending order based on perceived severity. Drawing on comments by study participants, it also indicates different participant groups that felt themselves threatened then or in the future and those most (but not exclusively) concerned about each category.

Higher up in the table are threats to fundamental human needs—survival, health and physical integrity, (cultural) identity, food and water availability—whereas economic and financial threats, whose expected or incurred costs can more readily be expressed in monetary terms, are listed toward the bottom of the table.

Table 5.1
Hierarchy of threat categories

Cost category	Threatened groups	Description	Associated participant groups
Existential or survival threat	Humanity, cultures, states, individuals	Climate change could lead to the destruction of the entire group (i.e., wipe out humanity, destroy cultures, lead to the disappearance of states or the death of individuals).	3-LH, 6-LL, Youth NGOs
Identity loss	Nations, cultures, communities, individuals	Identity loss can take many forms depending on the activities related to specific identities: the loss of an occupation (e.g., fishermen becoming farmers), the loss of homes, ritual sites, forms of cultural practice, landscapes, territory, seasons and possible experiences, species central for hunting, diet, experience of nature.	3-LH, 6-LL, Youth NGOs
Human suffering	Developing world, states, regions, communities, individuals	Human suffering refers to issues like poverty, hunger, hardship, injury, disease, and water scarcity.	1-HH, 5-ML, 6-LL
Extreme weather events	States, regions, communities, cities, individuals	Extreme weather events include storms, floods, droughts, heat waves, and the associated human, environmental, and economic losses, often mediated by the effects on agricultural productivity and infrastructure.	2-MH, 3-LH
Global food shortage	All humans, poor states and communities	This is a unique category in the sense that it is perceived to be a global concern, not linked to a particular place but systemic, worldwide. However, poor people are expected to suffer more from it.	4-HL

Table 5.1 (continued)

Cost category	Threatened groups	Description	Associated participant groups
Economic and development costs	States, regions, communities	This category refers to the loss of GDP due to the effects of climate change (overlap with extreme weather events) and the reversal of development progress (e.g., migration of fish populations leaving fishing communities and infrastructure stranded, or increasing temperatures decreasing agricultural yield or worker productivity).	1-HH, 2-MH, 3-LH, 6-LL
Economic costs of action	States, industries, voters	The financial costs and GDP loss associated with climate change policies; also the loss of global competitiveness.	4-HL, 5-ML

The literature on risk provides useful insights into the way people think about climate change–related threats (Slovic and Weber 2002). Risk perceptions are cognitive processes related to expected harm or costs. Three specific insights help deepen our understanding of the hierarchy of threats developed in table 5.1: (1) the distinction between expected threats to oneself and others, (2) the distinction between present and future threats, and (3) the quality or severity of the threats.

First, it matters whether individuals anticipate threats to themselves or others (Sjöberg 1998, 2000; Lorenzoni et al. 2006; Leiserowitz 2006). For many participant negotiators, the threat categories they believed applied to themselves (i.e., their states and peoples) and to others differed significantly. For example, island state representatives experienced an existential threat from climate change with regard to their own states, nations, and cultures, but anticipated only economic and financial threats for developed countries. Although all participants recognized threat categories that affected only other groups, they were naturally more concerned about threats to themselves. The participant groups associated with each category in table 5.1 tended to be those who connected the specific threat with their own groups, usually their own countries.

This distinction between self and others and between in- and out-groups already indicates a close relationship between threat perceptions and identity constructs. If you expect harm to come to a group you associate with, care about, or feel protective of, your motivation to act on climate change will be much stronger than if you believe that only other groups are threatened. Taking this argument one step further, individuals who are concerned about personal threats should have a stronger motivation to contribute to costly solutions than those who are not, especially if these are concerns about their personal survival and well-being.

The structure of the study participants' CAMs confirmed this pattern of differentiated threat perceptions to some extent. Members of groups 2-MH, 3-LH, and 6-LL linked expected climate change impacts to their own countries or identity groups. At the same time, they tended to be more concerned and demanded faster cooperative action than members of groups 4-HL and 5-ML. Although their CAMs had only weak connections between climate change impacts and their own countries, individuals in groups 4-HL and 5-ML generally recognized that climate change would strongly affect the poor and vulnerable in developing countries.

Second, participants distinguished between threats expected in the present and those expected in the future, which had implications for their sense of urgency. Individuals who had already experienced climate change–related damages in their home countries, and who lived in anticipation of additional damage every day had a stronger sense of urgency for action than those who believed that climate change impacts would occur in the future. Combining these distinctions between different threatened groups and the expected time of impacts, immediate threats to one's in-group appear to have a much stronger motivational potential than threats to one's out-groups in the future.

Closely related to the time dimension of climate threats is the effect of personal experience. According to work in psychology, "experientially derived knowledge is often more compelling and more likely to influence behavior than is abstract knowledge" (Epstein 1994, 711). The interview data confirmed this finding: participants who had observed and experienced what they considered to be climate change impacts (groups 1-HH, 2-MH, 3-LH, 6-LL) tended to have a much stronger sense of urgency than those who only had abstract or scientifically based conceptions of climate change–related threats (Nisbett and Ross 1980). Until recently, the dominant discourses in many developed countries framed climate change as a future problem. Due to an increasing number of extreme weather events, these discourses are currently shifting, with developed countries now conceiving of climate change as very much a present problem, one that is already affecting their communities today.

Although research on the two distinctions I explore above—affiliation with threatened groups and expected time of impacts—is expanding (Spence, Poortinga, and Pidgeon 2012; Safi, Smith, and Liu 2012; Bruneau, Dufour, and Saxe 2012), a third risk dimension has not yet received much attention in the literature. The threat hierarchy table suggests, however, that this third dimension is of fundamental importance for understanding differences in climate change–related beliefs: the quality of the expected threat.

Table 5.1 highlights that different study participants were concerned about different types of climate change–related threats. The interview transcripts revealed that individuals who were concerned about the three most severe threat categories—existential threats, identity loss, and human suffering—tended to use a normative decision-making and negotiation framework that was based on fundamental questions about right and wrong. Further, the three types of threats aroused very strong emotions, such as dread, fear, and anger, and were perceived to be so grave that allowing these impacts to occur was something these participants found morally unacceptable. The question in their minds was not whether these impacts could be avoided, or which ones should be avoided, but *how* they could be avoided.

The distinction between participants with a normative, rule-based decision-making framework and those with cognitive patterns focused on material costs bears strong similarities to the distinction made in moral philosophy between deontological and consequentialist reasoning. But the emotional pattern identified in study participants—strong and intense emotions among normative thinkers ("hot" deontology) and cooler, more subdued emotions among rational-choice thinkers ("cold" consequentialism)—is not congruent with prevailing moral theories, which consider deontology a logic- and rationality-based approach and consequentialism an approach driven by negative emotions like fear of pain (Greene 2008, 41).

The combination of cognitive content and emotion detectable in my study's data set supports recent and controversial work in social psychology and moral cognitive neuroscience, according to which deontological patterns of moral reasoning are driven by emotions, whereas consequentialist thinking—though not completely devoid of affect—is dominated by more calculating cognitive processes (Haidt 2001; Greene 2008). Offering yet another explanation for the observations made by Jonathan Haidt, Joshua Greene, and myself (Milkoreit 2014), Sabine Roeser (2010) argues that, ethical emotions and intuitions can be useful both as sources and as results of ethical reasoning and deliberation. Roeser takes issue with Daniel Kahneman's popular dual-process theory (2011), which distinguishes between a fast, affective system based on emotional heuristics and intuition (system 1) and a slow deliberative system based on rationality (system 2). Instead, she defends an integrated view of emotion and cognition and argues that emotions are part of both the fast and the slow cognitive systems, that "we cannot separate the cognitive from the affective aspect. They are two sides of the same coin. ... In the same vein it is futile to ask whether the affective or the cognitive response comes first" (Roeser 2010, 180). This integrated perspective is strongly aligned with Paul Thagard's theory of emotional coherence (2006, chap. 2; 2008), which I apply throughout this book.

The qualitative data gathered in my study are too general to indicate the direction of causal arrows (if any) between the three corners of the cognitive triangle. But they suggest a strong interaction between these three conceptual clusters. On the one hand, perceived threats to one's life, identity, or health or to that of one's group are associated with strong emotions and moral reasoning in categories of right and wrong (hot deontology). On the other hand, concepts related to "mere" economic and material loss for one's in-group are associated with weaker emotions and rationalist reasoning (cold consequentialism). These insights begin to illuminate how threat-related concepts impact other elements of a person's belief system, especially the overall motivation to act, the sense of urgency, and the applicable decision-making and negotiation frameworks.

Placing this debate in a broader context of decision making in social groups and conflicts, recent work in other disciplines confirms that a useful distinction can be made between two types of normative thinking—deontological and rationalist-consequentialist. Gregory Berns and colleagues

(2012, 754) argue that conflicts can emerge or escalate because individuals are not willing to trade off "sacred values, such as those associated with religious or ethnic identity" in return for material benefits. A value-based logic prevails over a material-consequentialist one (see also Atran and Axelrod 2008; Atran and Ginges 2012). Transferring this argument to the climate change context suggests that small island states are unlikely to be satisfied with financial and technological support for adaptation (although not even that form of support is readily forthcoming) when the survival of their statehood and cultures is at stake. Jeremy Ginges and Scott Atran (2011) make a similar argument about the deontological reasoning of those embroiled in violent conflicts. They contend that quantitative indicators or perceptions of the success or efficacy of violence are unimportant in the decision making of parties to the conflicts. These studies focus only on cognitive content, however, ignoring the question of emotions. Although Martin Hanselmann and Carmen Tanner (2008) begin to explore the role of negative emotions in decisions that involve sacred values, they fail to connect their analysis to morality.

Risk perception theory points to a number of additional factors that shape individual responses to climate change, including the notion of controllability (risks perceived to be under one's control are more acceptable than those controlled by others) and the distinction between more acceptable natural and less acceptable man-made risks (Sjöberg 2000). The CAMs analyzed here suggest that participant negotiators from developing countries (groups 2-MH, 3-LH, and 6-LL) had significantly heightened threat perceptions related to climate change because they lacked any control over the threat-creating factors (e.g., GHG emissions) and the means of mitigation. They even had limited control over the means of adaptation, which strongly constrained their sense of agency in the face of climate change. Regarding the distinction between man-made and natural threats, groups 1-HH, 2-MH, 3-LH, and 6-LL framed climate change as a man-made problem created by the developed world but increasingly exacerbated by the emerging economies. Groups 4-HL and 5-ML on the other hand, viewed climate change more as a natural threat or environmental management problem.

Based on these empirical observations, I disagree with Marco Grasso's assertion (2013, 377): "Moral cognitive neuroscience ... indicates that up, close and personal harm triggers deontological moral reasoning, whereas harm originating from impersonal moral violations, like those produced by climate change impacts, prompts consequentialist moral reasoning. Accordingly, climate ethics should be based on consequentialist approaches." Although the interview data confirm the argument that different types of expected harm trigger different normative frameworks of thinking that resemble deontology and consequentialism, the data also suggest that not all individuals involved in the climate change negotiations perceive climate change as an impersonal threat produced by nature rather than human beings. Even though study participants saw climate change as a phenomenon occurring in nature, many considered certain human beings—those in the developed world and the emerging economies—responsible for creating the threat and making it worse. What is more, because big emitters, by controlling all major emission sources, alone held the power to curb emissions, their failure to do so was morally reproachable and experienced in a very "up, close and personal" way by participants from less powerful and more vulnerable countries.

From the arguments outlined above, one might conclude that individuals from developed countries simply lack the heightened threat perception, intense emotional response, and urgency experienced by individuals from the developing world. Further, one might also conclude that this result is inevitable because it reflects the current science and reality of climate change—the developing countries will be hit harder and sooner than the developed parts of the world. But those conclusions, which amount to crude generalizations, ignore two important observations. First, as my study participants made clear, individuals in developed countries also experience serious threats related to climate change, threats linked to fundamental societal and individual values that define much of their identities—the neoliberal notions of individual freedom and choice, free markets and minimal government intervention, private property, and natural resource extraction as a source of prosperity and happiness. Taking action to mitigate climate change threatens to some extent all of these values in the way they are currently defined and practiced in the industrialized world, especially among conservatives in the United States. Because the perceived identity threat of such action to people in developed countries is similar to the perceived identity threat of climate change impacts to people in developing countries, it also elicits a strong defensive and emotional response. This response has, in turn, contributed to the phenomenon of climate change denial or skepticism, allowing strategic political and economic actors to exploit the cognitive conditions of their fellow citizens in developed countries, especially those with conservative ideological tendencies (Jacques, Dunlap, and Freeman 2008; Oreskes and Conway 2011; Jacques 2012).

Second, there were study participants from developed countries, in particular groups 4-HL, 5-ML, and various NGOs, whose cognitive patterns of concern, urgency, and normative reasoning closely resembled those of participants from vulnerable developing states. This phenomenon will become clearer when concepts of identity and norms are added to the analysis (see below). The general conclusion that different risk perceptions are insurmountable because they reflect the reality of the differing impacts of climate change around the world misses an important point: people's cognitive patterns can change based on perceptions of a changing environmental reality, personal experience (e.g., Hurricane Sandy, polar vortex events creating extreme winter conditions in the US Northeast, heat waves in Alaska, or a megadrought in the US Southwest), or other factors that increase the emotional intensity of responses to that reality (e.g., framing effects).

On another matter altogether, I observed a general and unexpected lack of concern among my study participants about environmental loss or degradation, extinction of species, and irreversible changes to ecosystems and landscapes. With very few exceptions, participants were not concerned about environmental values and threats to nonhuman life on Earth. When they mentioned these issues in their responses, they did so almost always instrumentally, referring to certain species, ecosystems, and environmental characteristics as necessary for the survival and well-being of humans. Not even participants from environmental NGOs showed a concern about the environment for its own sake, driving home the anthropocentric nature of the UNFCCC negotiations, where climate change is perceived not as an environmental issue, but as an inherently economic and social issue.

Other Observations Concerning Rational Choice and Cost-Benefit Analysis So far I have talked about costs as an important element of the rational-choice framework and therefore a key conceptual category for understanding political decision making. But what about other elements of the rational-choice framework? In this short detour, I offer some insights inferred from the study interview data.

The Absence of Benefits. In contrast to the breadth and specificity of their concepts regarding threats or expected costs of climate change, participants offered only a sparse, vague, and not very detailed set of concepts regarding the benefits of climate change policy. Examples included "to prevent catastrophic climate impacts," "to avoid dangerous climate change," and "green jobs." It is possible that climate change negotiators implicitly define the avoidance of climate change–related costs as benefits of action on climate change, but my study participants almost never said so explicitly. The paucity of their concepts about benefits reflects the negotiation reality, where the focus is on the need for mitigation, adaptation, finance, and technology support without addressing the expected benefits of any of these actions. If one assumes that decision makers employ cost-benefit analyses to choose a desirable path of action, this imbalance between costs and benefits in the minds of negotiators would present a problem, skewing the rational assessment toward the cost side of the equation.

Recently, the idea of co-benefits of action on climate change has become popular, which implies that acting on climate change has positive effects on non–climate change policy issues, including health, air quality, and security. This development supports the concern that the absence of perceived benefits of action on climate change is problematic for the negotiations. Co-benefits are supposed to fill that gap, complementing the avoidance of climate change–related costs.

The Absence of Numbers. Future costs and benefits of action on climate change need to be quantified for cost-benefit analyses in order to compare different policy options with one another. For example, a policy maker might want to know, in monetary terms, how much agricultural production loss related to climate change impacts can be avoided by investing a certain amount of money in mitigation, but such quantified comparisons do not seem to take place in the minds of policy makers. Indeed, they might not even be possible in a strict sense. With the exception of occasionally mentioned GDP losses associated with extreme weather events, not a single study participant quantified expected costs or benefits of different types of action on climate change, or different aspects of an international climate change agreement. It is understandably difficult to quantify various expected costs, including lost lives, extinction of species, and water shortages, or to find other measures that make these qualitatively different consequences comparable. My interviews and CAMs confirm this simple insight, challenging rational-choice theorists, who assume that cost-benefit calculations are at the core of all policy decisions.

In addition, a small number of participants (mainly in group 5-ML) recognized that cost-benefit analysis had severe limitations when it came to climate change. They argued that the tool had very specific areas of application, including national policy making and implementation, but it was not useful on its own to determine a country's national interest or shape its negotiation position. Other considerations, including reputational concerns, norms, and values, played a role that cost-benefit analysis was not able to capture.

Nevertheless, participants attempted to make rational decisions in the sense that they compared and weighed the likely but highly generalized consequences of different paths of action to the best of their knowledge and ability. In groups 4-HL and 5-ML, this often resulted in a general concern about the unspecified loss of international competitiveness related to mitigation policies and the associated political costs. Participants from developing countries tended to think in terms of increasing poverty, loss of agricultural yields, and water shortages, usually without attaching specific numbers, regions, or communities to these beliefs. Lacking any reliable numbers, they resorted to very rough qualitative comparisons, which suggests that the cost categories outlined above might be an important factor in climate change policy decision making. But even this generalized and numberless version of rationality seems to reach its limits when participants considered threats to humanity or to the international community, or the circumstance that sea-level rise would deprive a number of island states of the territory on which their statehood rested. The disappearance of states troubled the people of these states in ways that could not be usefully expressed in numbers.

Who Uses Cost-Benefit Analysis? Every study participant made reference to the idea of national interests and to the relevance of cost concerns related to action on climate change in the developed world. Although study participants seemed to intuitively understand and use the idea of cost-benefit analysis, they had three distinct approaches. Some applied this framework to their own situations, attempting to assess future costs and benefits for their own group. Others used it to mobilize action (e.g., co-benefit reframing attempts). Still others explicitly argued that cost-benefit analysis was a bad or unproductive way of looking at the world and often suggested that national interests were major obstacles in the negotiations or were immoral.

Collective Identities

No belief system can do without basic identity concepts that identify and distinguish existing actors and groups and that attach roles, characteristics, intentions, and other meanings to them. Identity groups are basic building blocks of belief systems—they determine which agents we associate with, seek to protect, assign blame or responsibility to, fight, fear or feel for. Identity concepts enable thinking about agency, group relationships, and norms. They are also preconditions for rational-choice thinking, because assessing costs and benefits is impossible without knowing who will incur or receive them. Rational cognitive patterns depend on identity—the existence of me and you, us and them, in-groups and out-groups (Tajfel 1982; Risse et al. 1999; Kaarbo 2003; Abdelal et al. 2006).

Identity Diversity—More Than the State or Nation For scholars of international politics, the natural starting point for thinking about identity is the state or the nation (Hall 1999; Risse et al. 1999; Kaarbo 2003). The interview transcripts revealed that nationality was rarely the only identity category, and sometimes not even the most important one, in the minds of participant negotiators. The following identity groups were the most common among study participants (see table 5.2): developed and developing countries, emerging economies (BASIC Group), big emitters, negotiation groups (e.g., AOSIS, LDCs), the vulnerable or poor in the developing world, humanity or human civilization, the planet Earth as a system of humans and ecosystems, oil lobbies, civil societies, and individuals.

These identity groups can be clustered into three general types. First, there were groups who used the state as the basic unit of membership. This included groups of countries (e.g., developed countries), process-specific groups of countries (e.g., UNFCCC negotiation groups), or issue-specific groups of countries (e.g., big emitters). Second, there were groups who used certain characteristics of their members to define the collective's boundaries. These groups tended to be larger than a state and to include the poor,

Table 5.2
Prevalence of identity groups

Identity group	Participant group*
National identity/The state	1-HH, 2-MH, 4-HL, 5-ML
Developed vs. developing countries, emerging economies	1-HH, 2-MH, 3-LH, 4-HL
Big emitters	4-HL, 5-ML, 6-LL
Humanity/Planet Earth	2-MH, 3-LH, 6-LL
The vulnerable/The poor	2-MH, 6-LL
Lobby groups, civil societies, individuals	All

*Not all members of listed participant groups shared the norms opposite their listings.

the vulnerable, and even all of humanity. Finally, there were groups whose members used their association with domestic or local politics, including lobby groups, civil society organizations, or individual political actors to define their collective identity.

Truly extraordinary was some participants' identification of planet Earth as an interconnected system of humans and all other life, as a unit they could identify with and seek to protect. Short of thinking about other civilizations in the universe, this is the ultimate, all-encompassing in-group, extending the boundaries of what is worth protecting even beyond humanity, animal, and plant life to the systemic functions of the planet itself.

Some of the categories participants used present a troubling puzzle for international relations theory. The modern international system and those who study it only acknowledge states as actors with formal rights and obligations. Sovereignty is a state-specific property, which grants a state the right to represent its citizens, to enter into agreements with other states, and to create international regimes (Keohane 1984; Koremenos, Lipson, and Snidal 2003), to conduct diplomacy, to appoint officials of international organizations, and to subject itself to rules on the conduct of war (Held 1995, 78). Although this system also acknowledges the existence and perhaps even influence of other actors, including nongovernmental organizations, transnational networks (Keck and Sikkink 1998), and private sector organizations (Keohane and Nye 1977; Strange 1996), it does not award these nonstate actors any rights or obligations beyond those they are given by the states in which they originate. And even though the Paris Agreement explicitly acknowledges nonstate actors and their role in achieving the goals of the UNFCCC, this acknowledgment neither confers any rights on nonstate actors nor gives them formal standing as potential parties to the UNFCCC. It merely serves to encourage them to maintain or step up their involvement in the global climate governance process, especially their contributions to GHG emission reductions and renewable energy transitions of companies, communities, cities, and regions.

Given this background, the frequent reference to nonstate actors by participants, especially by participant diplomats, deserves attention. These references are unproblematic when they merely describe current or desirable behavior by nonstate actors, who play a certain role in the larger context of global climate governance. But references that attribute rights, obligations, or both to humanity, "the wealthy," or "the vulnerable" are deeply problematic to the extent that they imply a fundamental mismatch between the actor categories accepted in the current international system and the actor categories necessary to deal with the climate change problem. Indeed, they pose an ontological challenge by attributing rights and obligations to entities that have no standing in international politics. Nobody is able or entitled to represent humanity or the poor in the international system—these entities are simply abstractions as far as the institutions of world politics are concerned. Appeals to them are bound to be ineffective, yet climate change negotiators (not only participant negotiators) make such appeals and assign these abstract entities rights and obligations that appear to be rooted in cosmopolitan notions of mutual responsibility based on shared humanity.

How can one understand these cognitive patterns? One could argue that powerless states invoke the protection of humanity and our moral obligations toward it for purely instrumental reasons—moral persuasion is the only form of power they can wield given their lack of material power or mitigation potential. But, in many cases, it is equally plausible that the climate change problem is simply incomprehensible without reference to humanity as a whole—the concept creates emotional coherence in response to conceptual novelties regarding the global climate system and the complex, uncertain, long-term consequences of climate change. Once a person discovers the interconnectedness of all human beings through the global climate system, it becomes easier or even imperative to believe that the broader collective interest of all ought to trump the narrow national interests of a few. People who share this belief are also more likely to question the adequacy of the current international system to address climate change. Rather than misunderstanding the basic rules of the system or being hopeless moral optimists, they have acquired beliefs that are incomprehensible within or even incommensurable with the reality of the UN process. They begin to challenge the system in order to reestablish cognitive coherence.

Underestimated: Place Identity An identity concept receiving greater attention outside the field of international relations (Proshansky, Fabian, and Kaminoff 1983; Dirlik 1999; Adger et al. 2011; Fresque-Baxter and Armitage 2012), place identity is defined as "those dimensions of the self that define an individual's personal identity in relation to the physical environment by means of a complex pattern of conscious and unconscious ideas, beliefs, preferences, feelings, values, goals and behavioral tendencies" (Proshansky 1978, 155). Its many elements include emotional attachment, continuity, security, and social connections (Fresque-Baxter and Armitage 2012). Naturally, place identity is of major significance for those who perceive the physical integrity of their home countries or regions to be at risk from climate change. Among my study participants, this diverse group of those so affected was dominated by representatives of small-island and low-lying developing states in groups 3-LH and 6-LL. For them, a key part of their identity—their place identity—was existentially threatened. In a few decades, places they held dear might disappear beneath the rising ocean, or be made unrecognizable by the ravages of coastal flooding, erosion, and loss of land. The impacts of climate change might destroy places they called home, where they grew up, where they used to work, where they met their spouses. They saw climate change as threatening to completely change the topography of their world and the world of their children and future generations, taking away physical realities that linked their present to their past.

But it was not only islanders and lowlanders who experienced threats to and even loss of place identity. Study participants in groups 1-HH (youth NGOs) and 4-HL expressed concerns about the loss or dramatic change of seasons where they were from, for example, the absence of snow in some regions in North America in the winter of 2011–2012, the loss of local forests, the sadness and melancholy associated with losing a way of life, "the world as it always was." With one exception, all of them were women, for whom nature and beauty were a central part of their lives. Their place identity had to do with a deep attachment, not to a country or nation, but to a region, a state or province, a city, a cultural community, or an ecosystem with which they felt a special connection.

To those whose place identity is threatened, the world feels less certain, less reliable, and less safe. The mental costs of accepting and adjusting to a changing environment, especially when it is unclear when the changes will end and what the new environment will look like, can be immense. These changes threaten what Anthony Giddens (1991, chap. 2) calls a person's ontological security—a sense of stability and continuity of the basic constituent elements of life.

Societies tend to develop a set of tools or sociological mechanisms to deal with—or more often suppress—such disturbing information or signs of impending change (Norgaard 2011; Weintrobe 2012). One would expect that climate change negotiators are forced by their very profession to confront these disturbing signs. Yet, most of the time, such signs, whether in the form of scientific findings or individual accounts, are kept away from the negotiation process. The NGO community tries time and again to introduce these elements into the climate change conversation, but their side events or staged protests are kept away from the negotiations, often taking place in a different building or locale, and negotiators rarely attend them. There is, however, one UNFCCC negotiation group who persists in directly introducing disturbing accounts of climate change impacts and their tragic human consequences into the actual negotiation space—the Alliance of Small Island States (AOSIS), whose delegates emphasize the need to devise solutions based on the latest science.

The examples presented above were all offered by study participants, who perceived threats to the place identity of themselves or their in-groups. But could people experience similar emotions from the loss of places that are *not* their own, for example, the "loss of paradise" for world-traveling Europeans with the disappearance of small-island states in the South Pacific, or the "loss of mystery" with the changing Arctic, and associated notions of adventure, courage, uniqueness, beauty, and value? And could they do so from the loss of a place they had never visited or never intended to see?

Notions of Justice

When talking about threat perceptions—the first corner of the cognitive triangle—I mentioned that some are linked to deontological thinking (moral questions of right and wrong), whereas others are linked to consequentialist thinking (maximizing the positive effects of an action). But that distinction left open important questions concerning the substance of these moral norms or positive effects. What determines right and wrong? How should positive be assessed?

Norm Content and Advocates   Who cares about norms, what are they, and whom do they burden with obligations? With the exception of group 4-HL and, surprisingly, some representatives of faith and development NGOs, all participant groups integrated normative beliefs into their cognitive structure. Table 5.3 summarizes the set of normative ideas regarding global climate governance contained in the interview data. Most of these were shared across multiple groups.

Table 5.3
Norm diversity and distribution

Norm	Participant group*
Every state should contribute/care/take a fair share of the burden/based on its capabilities	1-HH, 2-MH, 3-LH, 5-ML, 6-LL
Historical responsibility	2-MH
Richer countries should do more	2-MH, 3-LH, 5-ML, 6-LL
Emerging economies should do more/More integrated mitigation.	2-MH, 3-LH
Polluter pays principle	1-HH
Solidarity/Protection of the weakest/Pro-poor regime/Equity favoring the poor/The rich help the poor	1-HH, 3-LH, 6-LL
Human responsibility to other human beings	4-HL, 5-ML, 6-LL
Global wealth equalization through resource flows from North to South (equity)	2-MH
Multilateralism and inclusiveness/Procedural justice	1-HH, 2-MH
Intergenerational responsibility	2-MH, 5-ML
Changing lifestyle patterns in the rich world	1-HH, 2-MH, 3-LH, 6-LL

*Not all members of listed participant groups shared the norms opposite their listings.

Most of these norms do not contradict or mutually exclude one another, at least not in the general formulations of the participants. That said, the norms mentioned by study participants are vague and provide no clear guidance for the distribution of responsibilities or burden-sharing arrangements among negotiation parties. They merely suggest that every country should contribute in some form, that countries' contributions can differ, and that developed countries ("the better-off," "the rich") and, in some participants' opinion, the emerging economies should do more than they were doing then. Let me focus briefly on a few of the most contentious norms in the global climate change negotiations: historical responsibility and Common but Differentiated Responsibilities and Respective Capabilities (CBDR-RC).

Only two out of six members of group 2-MH supported the concept of historical responsibility. Individuals in group 4-HL considered it unproductive or associated it with negative emotions. Most participants did not even mention the concepts of equity and historical responsibility, which have always been a major source of contention in the climate change negotiations. But many study participants believed that fairness and responsibility were important concepts. Finding a productive bridge between these two levels of norms—equity and historical responsibility as a position in the negotiations, on the one hand, and a subjective preference for fairness and responsibility, on the other—might be an excellent opportunity to advance the negotiations. Although the Paris Agreement makes no mention of historical responsibility (which, one can hope, may open up new and less contentious normative space for future regime development), it does mention Common but Differentiated Responsibilities and Respective Capabilities (CBDR-RC) and introduces differentiated responsibilities in multiple places.

The idea that all states—even developing states—should contribute to a solution to the best of their respective abilities might not seem surprising, but it signals a significant departure from the previously dominant interpretation of the CBDR-RC, a core principle of the UNFCCC. In the past, this norm has been operationalized through the distinction between developed countries in Annex 1 of the Kyoto Protocol and non–Annex 1 countries. The Kyoto Protocol required Annex 1 countries to take on legally binding mitigation commitments and to provide resources for mitigation and adaptation action in non–Annex 1 countries, which had no legally binding mitigation obligations under the protocol. Since the 2012 COP in Doha, however, these countries have been asked to take Nationally Appropriate Mitigation Actions (NAMAs). In 2012, almost all study participants considered the distinction between Annex 1 and non–Annex 1 countries unproductive and even an obstacle to successful negotiations. Returning to the idea that there are common rather than differentiated responsibilities, study participants supported the idea that every country, regardless of its development status, should contribute to a climate change solution. In some CAMs, especially those of group 3-LH, this norm was connected to the emerging economies, who were not part of Annex 1, but who no longer fit the category of developing countries because of their increasing economic success, emissions, mitigation potential, and global political power.

This trend toward establishing similar responsibilities for all parties to the UNFCCC negotiations was reflected in the negotiation frameworks developed between the 2011 (Durban) and 2015 (Paris) COPs. The Durban Platform for Enhanced Action established a new subsidiary body of the UNFCCC with the mandate "to develop a protocol, another legal instrument or an agreed outcome with legal force under the Convention applicable to all Parties" (COP Decision 1/CP.17), launching the process that led to the Paris Agreement. The new terminology for states' obligations under the future agreement focused on intended nationally determined contributions (INDCs)—expressions of intended actions to be formulated bottom-up by each country based on its own assessments of its capabilities and adequate ("equitable") contribution to the global public good. Countries were encouraged to submit their INDCs ahead of the Paris negotiations in order to allow for peer review and potential adjustments. All parties to the negotiations, not only those considered developed countries, were expected to submit INDCs.

Building on the idea of universal, but voluntary contributions, the Paris Agreement did away with the Kyoto "firewall" between developed and developing countries and established similar kinds of obligations for all states, regardless of their development status. Every party to the Paris Agreement is required to contribute to the shared global goals, which now include a temperature target ("well below 2°C"), an increase in adaptation abilities and climate resilience and in the consistency of finance flows with "a pathway towards low greenhouse gas emissions and climate-resilient development" (Article 2.1 PA). As the price for this universal applicability of obligations under the Paris Agreement, the parties' minimum contributions are neither determined nor legally binding. Instead: "Each Party shall prepare, communicate and maintain successive nationally determined contributions that it intends to achieve" (Article 4.2 PA); in other words, each country merely has to present promises of intended future actions. These NDC packages of promised actions are supposed to be revised every five years and reviewed with a set of transparency and stocktaking mechanisms laid out in Articles 13–15 of the Paris Agreement. The regular review cycles become the central tool for measuring progress toward the global goal and for incentivizing additional (also voluntary) actions in case this progress falls short, however that would be determined.

The hope is that this new hybrid model of global climate governance, which combines bottom-up action and top-down review, provides positive incentives and support but no legal requirements and will lead parties to increase their contributions over time. The psychology of constructive shaming and peer pressure is supposed to replace the psychology of international legal commitments, which was at the heart of the Kyoto Protocol (and many other multilateral environmental agreements before that) and which is now widely considered an unworkable model for international environmental treaty making in the twenty-first century.

Detour: Some Notes on Social Constructivism and the Relevance of Ideas The CAMs developed for this study offer an example for the relevance of a purely ideational construct for an ongoing political process: the argument that solving climate change requires a "new model of development" or "a new aspirational model for the poor."

Members of three different participant groups (1-HH, 2-MH, and 4-HL) formulated different versions of the same argument: apart from the various technological, economic, legal, and institutional changes discussed in the negotiations, addressing climate change would require a new development paradigm. The demand for a new model of development is often associated with a criticism of current consumption patterns in the world, and the pursuit of economic growth as the ultimate political and social good. However, the participants' argument did not focus on the behaviors, policies, practices, or outcomes of development, but on the ideational construct or conceptual model that defined, gave meaning to, and justified what policy makers today understand as development. A model in the sense used by the study participants is an abstraction—a cognitive structure. Although not physical itself, it relates to a broad range of physical objects, as well as to the physical-material realities of human behaviors, institutions, and practices, including patterns of production and consumption, economic growth, professions, industry standards, goods and services, global supply chains, and social relationships. Study participants who argued for a different model identified some of these current physical-material realities of development as problems in the context of climate change. Seeking to permanently change these realities and patterns of behavior, for example, the kind and amount of consumption in the developed world or the pursuit of economic growth at the expense of other goals, they suggested that such change required, first of all, a change of beliefs, "a change of mind," presumably by everyone involved in the practice of development. Since a broad set of undesirable economic and political behaviors today rests on a model of development that prescribes these existing patterns and allows for their institutionalization, to tackle these behaviors, we must first tackle the ideational foundation that creates and reproduces them.

The importance of ideas in changing physical-material realities is a feature of Robert Cox's neo-Gramscian theory of hegemony and acquiescence in the international system. According to Cox (1996), hegemony is enabled by an alignment of ideational, institutional, and material power. In the present international system, neoliberalism provides the foundation for the institutional architecture and justifies the distribution and accumulation of material capabilities and power in certain parts of the system. Demanding a new development paradigm strikes at the ideational foundations of the current international system, with significant implications for its institutional and material components. Steven Bernstein (2001, chap. 5) makes a similar argument about the ideational power of neoliberalism in the context of global environmental governance when he suggests that this particular set of ideas and norms has been so successful because of its socio-evolutionary fitness both in the existing social structure and with the actors at the time of its rise.

Regardless of the validity of Bernstein's argument, designing a new development model and creating the conditions for its broad acceptance and application are not tasks that can be accomplished by climate change negotiators. If these study participants have correctly identified a problem and corresponding solution space outside the negotiations, this would have important implications for the UNFCCC process. It might imply that the negotiations are not able to succeed because not all pieces of the puzzle are in the possession of the negotiation parties. Study participants who made the new-paradigm argument did not draw this conclusion for the negotiations, however, nor did they elaborate on its implications for climate or development governance.

Linking the Corners of the Triangle

Now that we have explored the cognitive diversity at all three corners of the triangle, what are the patterns that connect threat, identity, and justice concepts? Linking norm-related concepts to threat and identity concepts is not straightforward. But, turning to the study data set, if one combines the participants' threat hierarchy (table 5.1), identity constructs (table 5.2), and the norms summarized in table 5.3, two patterns emerge. To better see them, it is useful to think about each corner of the cognitive triangle as having two distinct settings, separated by a threshold.

First, there is a threshold separating two major types of threats: those concerning human life and well-being, including existential threats, identity loss, and human suffering, and those of a more economic or financial character. Participants who were concerned about threats of the first type, that is, threats above a certain severity threshold, tended to use a deontological framework when considering the obligations of states to act on climate change; for them, the negotiations were about the right or wrong path of action. In contrast, participants who cared mainly about threats of the second type, that is, below the severity threshold, such as economic losses due to higher energy costs or infrastructure damage, tended to use a consequentialist framework; for them, the negotiations were about the most cost-effective or efficient path of action. The threats triggering deontological reasoning among participants were all related to direct climate change impacts, whereas those triggering participants' consequentialist arguments were related to policy and action on climate change. This distribution split the participants into those who thought mainly about the negative effects of inaction (groups 2-MH, 3-LH, and 6-LL, youth, faith, environmental and development NGOs) and those who were focused on the negative effects of action (mainly groups 4-HL and 5-ML). Group 1-HH is interesting because it occupied the middle ground between the two camps. This observation suggests that fundamentally different ways of normative thinking are associated with different cost categories.

Threat perceptions are not only linked to people's moral judgments, but also to their mental representations of their collective identities. Threats have to relate to a group in order to be perceived as threats; they must have a target to be perceivable. Without such a link between the phenomenon in question and a threatened group, it would not even be thought of as a threat. The target that tends to be most important is one's in-group—the people one identifies with, shares a history with, and seeks to protect.

As mentioned above, the interviews revealed a much larger set of identity constructs than one would expect in an international negotiation setting. Naturally, states and nations are the most important identity group in such a setting. But other identity categories sometimes complemented or even competed with the national identity of participant negotiators.

As a general rule, the larger (more inclusive) a participant's in-group, the more likely that participant was to focus on threats above the severity threshold. Inclusivity here refers to the diversity, number, and geographic location of in-group members. Examples for large, inclusive identities included humanity, all vulnerable people, or all developing countries. The less inclusive a participant's identity constructs were (e.g., limited to a single nation or state), the more likely the participant would be concerned about threat types below the severity threshold; such participants would assess the largest good for the largest number in a purely domestic rather than global context.

There are important exceptions to this observation. Participants representing highly vulnerable countries only needed a local or national in-group definition to make use of a deontological framework, considering action on climate change as simply the right thing to do. The severe, sometimes existential threats of climate change to their respective communities or nations, combined with their inability to fight these threats alone, easily resulted in beliefs that involved a shared humanity and the duties of the powerful to help the powerless. Participants from countries with low vulnerability to climate change, on the other hand, needed a more cosmopolitan in-group definition to have similar threat perceptions and desires for multilateral cooperation. Many NGO representatives from developed countries, especially those in youth and development organizations, held such inclusive identity beliefs, identifying, for example, with "every person under 25 on the planet" or with "the poor in the developing world." But there were also a number of participant diplomats with large and inclusive identities. The participants' belief patterns thus demonstrated that negotiators from different regions of the world viewed the climate change problem through very different moral lenses, depending on their idiosyncratic threat perceptions, which were driven by their specific collective identities.

The two emerging patterns can therefore be summarized as

• Hot concern (figure 5.2): Severe threat perceptions, highly inclusive identities, and deontological moral thinking
• Cold calculation (figure 5.3): Less severe threat perceptions, less inclusive identities, and consequentialist moral thinking

Distinguishing two distinct settings for each corner of the cognitive triangle, though somewhat artificial, is very useful when talking about the participants' cognitive-emotional patterns. Each corner might be more appropriately understood as containing a range of ideas, but also a threshold where the character of the cognitive elements changes qualitatively. This is a bit like the burner of your electric stove, which has a low, medium, and high setting. When you turn the dial up, at some point along the way

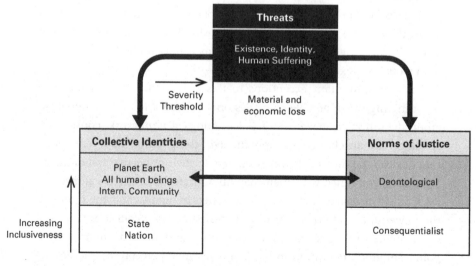

Figure 5.2
Cognitive triangle setting 1: Hot concern.

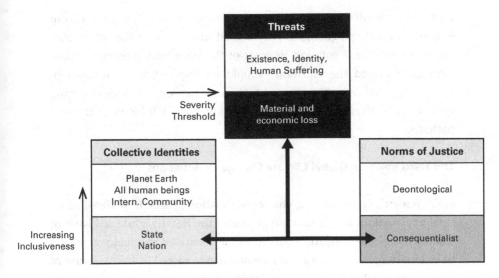

Figure 5.3
Cognitive triangle setting 2: Cold calculation.

the heat level switches from what you would describe as low to high. The same principle operates here when distinguishing between monetary and human needs–related threats, inclusive and exclusive identities, and deontological and consequentialist ethical norms.

How did these patterns play out among the study participants? Members of groups 3-LH and 6-LL strongly displayed patterns of hot concern. They perceived present threats to their existence and identity, often used an explicit moral framework, and expressed support for norms that invoked humanity and solidarity as reference points. These norms obligated everybody—every state and every human being—to take responsibility for others, especially the vulnerable. Groups 2-MH and 5-ML showed a similar pattern of normative thinking, and expanded it to a responsibility to future generations.

Group 4-HL was most clearly associated with cold calculation, identifying quantifiable threats, often remaining rooted in a national identity, and approaching decisions using a consequentialist framework based on rational choice and utility maximization. The arguments presented in this group rarely seemed to contain any normative statements.

Members of group 1-HH, who represented countries that were both highly vulnerable to climate change impacts but also increasingly significant contributors to global GHG emissions, demonstrated a form of compassionate pragmatism. They argued that everyone should contribute to a multilateral solution in favor of the poor, and, preferably, those who have caused the emissions should be responsible for mitigation. They recognized that this group of high emitters would include emerging economies in the future. Members of group 1-HH fell between the two patterns.

The Third Image in Global Climate Change Negotiations

Much research on the climate change negotiation process has offered structural explanations for its slow progress and the heavily contested issue of sharing the global mitigation burden. It has identified two major obstacles to international cooperation on climate change rooted in the structure of the international system: power and vulnerability imbalances between poor and rich countries; and the configuration of the global economy, which shapes various actors' interests and distributes the resources to defend these interests in ways that result in noncooperative outcomes. Below I explore how those engaged in the climate change negotiations think about the structure of the international system and how it affects climate politics in the UNFCCC.

Global Power Structures

Without being specifically asked about this issue, members of all six participant groups described global power structures and their relevance for the climate change negotiations at some point during their interviews. They made three distinct arguments, basing them on (1) the economic interests of individual states; (2) the economic interests of vested interest groups at the domestic level; and (3) the relationship between the developed and the emerging economies.

First, some members of groups 1-HH, 3-LH, 4-HL, and 5-ML believed that states' concerns about competitiveness at the international level were a major driver of the negotiation dynamics. Although they never elaborated on the concept, they implied that action on climate change imposed unequal costs on parties to international economic transactions. A country's expected economic disadvantages in comparison to other countries would be detrimental to its competitiveness, resulting in economic damages. Participants never mentioned specific kinds of damages, but presumably these would include a reduction in GDP and the loss of both jobs and taxable income. Most study participants believed that the expected cost of these damages was the main reason why the developed and emerging economies were not willing to act decisively on climate change. This argument was associated with trade-offs between the present and the future, "carbon leakage," the current structure of major economies, the energy efficiency of current industries, stranded assets, and the idea of keeping a level playing field for all global economic players. The loss-of-competitiveness concerns strongly linked the climate change issue to global economic governance. In a broader sense, theirs was a system-level argument concerning the balance of economic power in the international trade system: countries that were in a powerful and advantaged position wanted to stay ahead and to avoid strengthening others at their own expense, for example, by providing finance and technology to their economic competitors.

Second, some participants, especially in groups 3-LH, 4-HL and 6-LL, argued that vested interests—oil, coal, and gas interests in particular countries—played an important role in climate change politics because they had substantial financial resources and easy access to the political system. Presumably, they used their money and connections to prevent domestic action on climate change, indirectly hampering progress in the international negotiations. Although the competitiveness argument outlined above suggests that action to mitigate climate change might be desirable and possible if a level economic playing field could be maintained, participants believed that vested interests sought to preserve the status quo in order to protect their own profitability and expected income flows in a business-as-usual future.

Third, some participants, especially in the Environment & Market NGO group and in group 3-LH, argued that the world was undergoing a major power transition, with economic power shifting from the developed to the emerging economies. This presented a significant constraint for the climate change negotiations because it exacerbated the competitiveness concerns of both of these groups. Already seeing their power waning, the developed economies were not interested in speeding up this transition by imposing the costs of action on climate change on the citizens of their own countries. The most negative interpretation of this argument even suggests that the rich had no interest in fighting climate change because its impacts would limit or roll back the progress made elsewhere in the world. The emerging economies were not interested in curbing emissions at the point when they were finally catching up with the developed world and were relying on economic growth to maintain social stability at home. According to this argument, climate change diplomacy was an instrument of power politics— managing systemic changes that were perceived as threats or opportunities by different state actors.

The Two-Level Game

Every participant group acknowledged the relevance of cross-scale link- ages and the influence of domestic politics on global climate governance, which Robert Putnam (1988, 433) characterized as a two-level game: When national leaders must win ratification from their constituents for an inter- national agreement, their negotiating behavior reflects the simultaneous imperatives of both a domestic political game and an international game. The interview transcripts present strong evidence that the case of the United States and the experience of the Kyoto Protocol brought this issue into sharp focus for study participants. Whenever they made arguments related to domestic politics, for example, discussing a lack of public sup- port, the role of vested interests, political institutions, and ideologies, these were almost always connected to the United States, where participants saw only weak public support for climate change policy, where Congress was opposed to binding legislation or treaties on mitigating climate change, where vested interests were strong and had significant influence in the political system, and where a conservative ideology had led to widespread climate change denial and skepticism (Inhofe 2012).

Indeed, the massive mitigation potential of the United States, its leader- ship legacy regarding global public goods, and its identity ties to capital- ism, unbounded material success, and excessive consumerism have played a unique role in the global climate change negotiations. And the attention paid to the domestic politics in the United States and how they relate to the UNFCCC process has also been unique; in the lead-up to both the 2009 COP in Copenhagen and the 2015 COP in Paris, virtually every UNFCCC negotiator was aware of the fact that the United States was not able to make legally binding international mitigation commitments because Congress would simply not ratify any agreement that called for them. This was the key lesson learned from the fate of the Kyoto Protocol.

Before Paris, one could have argued that the failure of global climate governance was a clear case of regime failure without the hegemon: the climate change case demonstrated the necessity of having a proactive, regime-supporting stance of the most powerful actor in the system—the United States in the view of almost all the negotiators. Although Robert Falkner (2006) argued that it was not always necessary for the hegemon to support a regime-building effort, to coerce and incentivize others into a new set of behaviors, and that environmental regimes could succeed even *against* the will of the hegemon, in the case of climate change, an unwilling hegemon appeared to be a serious obstacle. Here the global public good simply could not be provided without the hegemon pulling its weight. At least it was the dominant perception among negotiators of developing countries, including the hegemonic runner-up China, before Paris, that the United States had to act first. This belief in the need for hegemonic leadership created widespread concerns about regime effectiveness and ultimately deterred many negotiation parties from making their own commitments.

After President George W. Bush refused to sign the Kyoto Protocol in 2001, claiming that its adoption "would have wrecked our economy," and after the United States (along with the BASIC countries) ensured that the mitigation commitments of the Copenhagen Accord in 2009 would not be legally binding, most study participants viewed the US domestic situation as a severe constraint on meaningful climate change negotiations, even "holding the negotiations hostage." Some participants seemed to go one step further, suggesting that specific national institutions—democracy and short election cycles—posed challenges to substantial climate change negotiations because they reinforced and institutionalized short-term thinking and political incentives. This raises the intriguing question whether democracy is unfit for global governance. More generally, the time period since the adoption of the Kyoto Protocol was one of growing pessimism about the ability of multilateralism and diplomacy to address global challenges. Effective multilateral cooperation seemed to be increasingly elusive in a changing global context.

At the same time, many participants considered the two-level game insight helpful, identifying barriers to success that could be addressed, although not always within the scope of the negotiations. Some participants proposed strategies to address these barriers, for example, mobilizing the public in the United States and working with its civil society and the media to devise and spread appropriate messages at the grassroots level. Refocusing on the role of voters, these participants believed that pressure from below could counter vested interests and create the political will for climate legislation. Recognizing that the enabling conditions for an international agreement could not be created at the level of political elites, these ideas opened up new solution space outside the negotiations.

The Paris Agreement, at least to some extent, vindicated both the hegemonic leadership requirement and the two-level game theory. One of the most important changes between the 2009 and 2015 COPs that facilitated the agreement was the development of a strong bilateral relationship between the United States and China with concrete collaboration projects on climate change and renewable energy. In November 2014, the presidents of the two biggest emitters made a joint statement on climate change, affirming their commitment to both the UNFCCC process and domestic action. Both countries also made substantial headway on setting caps to domestic emissions, indicating their serious intention to act and not just talk. Once both the hegemon and the hegemon-to-be stepped into their leadership roles, real progress became possible.

Also, recognizing the continuing problem of unachievable congressional approval in the United States, many UNFCCC delegations were prepared to accept a kind of deal that would not require such approval, which meant that the Paris Agreement would not establish legally binding emission reductions or climate finance commitments (Bodansky 2016). In essence, US domestic politics—the obstinacy of Congress—pushed the international community toward the new hybrid model of global governance, replacing legally binding (top-down) commitments with voluntary (bottom-up) promises of action. Necessity became the mother of invention: the pledge and review system.

Realism's Eternal Wisdom?

The observed cognitive patterns of study participants—a focus on the costs of action, major differences in the expected types of costs for different actors, and their timing—mirror the current physical-material circumstances and scientific information about climate change. Each belief system is a reflection and interpretation of reality or expectation of future developments.

Historically, the developed world has been and continues to be the main contributor to climate change. The expected impacts of climate change will disproportionately affect the developing world—poor and vulnerable countries with limited resources to adapt to change. These negative impacts are an unintended consequence of economic activity in the industrialized countries, which alone have the power to address the problem. Only they can effectively control the emission sources, and only they have the resources needed to help the poor and vulnerable countries deal with the climate change impacts to come. Effectively dealing with the problem requires at a minimum the costly transformation of current economic structures, phasing out fossil fuels from all production and consumption processes. This includes a teardown of some of the most profitable industries on the planet, but it affects all sectors of the global economy. Power, defined here as the capacity to deal with climate change, is unevenly distributed in the world; it is concentrated in the global North, though this power is currently expanding to the emerging economies. The powerful actors' current perceptions of their material-economic interests—preserving economic power and protecting profit-making industries—do not favor action to mitigate climate change because mitigation implies economic costs that are not outweighed by identifiable benefits. Consequently, no action is taken to slow climate change, condemning the poor and vulnerable to increasingly certain damages and suffering.

In short, the distribution of power and threat perceptions over the last two decades has resulted in political behavior that confirms the expectations of neorealist theory. The powerful act in their self-interest to the detriment of the powerless. The Paris Agreement has not really changed this assessment. It requires neither decisive action on the part of the rich nor significant support for the poor. Whether the bottom-up approach to action on climate change will deliver results over time remains to be seen. Although all study participants—men and women alike—shared this realist understanding of the climate change negotiation context, they differed in their beliefs regarding the inevitability of noncooperation. Those differences will become clearer when analyzing notions of identity, norms, and models of change below.

Does Science Matter When You Negotiate Climate Change?

Policy making in modern societies is based on the general assumption that scientific knowledge is a necessary ingredient for good decisions (Norgaard 2009). Making a distinction between science as the pursuit of truth and politics as a struggle between values, many scholars, policy makers, and voters assume that there is a logical link between a scientific finding, such as anthropogenic climate change, and the "right" policy response. Mike Hulme (2009, 102–103) calls this model of the science-policy interface technocracy.

Given the nature of climate change as an environmental phenomenon rooted in physical, chemical, and biological processes, scientific knowledge should play a major role in the way negotiators think about climate governance and climate change policy. Science-based concepts should determine or at least influence negotiators' understanding of the problem, possible solution options, the overall goal of global climate governance, and perhaps even the social and technological challenges to be expected in pursuit of this goal. The Intergovernmental Panel on Climate Change (IPCC) acts as the key source for scientific information in the UNFCCC process. Its task is to aggregate and communicate all relevant scientific information for regime-building purposes, that is, to be policy relevant without making policy prescriptions, a task reserved for the politically legitimized representatives of the states that are parties to the negotiation process. In this ideal world, scientists serve up the truth to the best of their collective current knowledge, and diplomats decide what to do about the factual state of the world. The real world turns out to work very differently.

The interview data disappoint any expectations about the relevance and use of climate change science and illuminate the messy nature of science-policy interactions. First, and somewhat disturbingly, science did not play a significant role in the belief systems of most study participants. Rather than integrating science into their various thought patterns on the nature of the climate change problem, solution options, and the global climate governance goal, the participants kept their scientific knowledge almost entirely separate from them. The minds of study participants contained a set of scientific concepts that could be activated upon questioning, but that did not necessarily interact with their thought processes regarding the negotiation process. Participant diplomats used neither scientific concepts in general nor complexity concepts in particular to define their negotiation positions, solution options, or the overarching goal of global climate governance.

The four subsections below explore distinct science-related cognitive patterns that emerged from the participants' cognitive-affective maps: (1) a general disconnect between science and politics; (2) beliefs concerning the ultimate goal of global climate governance; (3) the absence of complex-systems thinking; and (4) beliefs related to time. These findings confirm that some, but not all, complexity-related characteristics of climate change pose particular cognitive challenges for those engaged in the climate change negotiations.

The Science-Policy Disconnect

All study participants shared a similar set of ideas with respect to the nature of the problem they were supposed to address. This belief cluster is rooted in science and usually has four components: climate change is real, anthropogenic, caused mainly by GHG emissions and land-use change, and has a range of serious consequences. This quartet of concepts has become a background truth for all political thinking about climate change without making reference to numbers, uncertainties, timelines, or other specifics. It is taken for granted, does not seem to require articulation anymore, and appears to be a sufficient basis for debating action on climate change. Most participants were satisfied with this simplified summary of the challenge at hand—GHG emissions have serious consequences—as a cognitive framework that provided direction. At the same time, these participants struggled to define success for the climate governance regime. They did not know how to measure success, what would constitute useful milestones, and what relevant time constraints might exist.

The participants' cognitive-affective maps demonstrate this disconnect with an absence rather than presence of links between certain concepts. Across all CAMs, there were only two consistent links between scientific and political concepts. The first was a strong link between GHG emissions, a scientific concept, and mitigation as a solution- or action-oriented concept. The second was between multiple scientific concepts (e.g., sea-level rise, temperature change) and concepts of expected climate change impacts (e.g., coastal erosion, disappearance of island states, failed crops).

The disconnect between the political aspects of the negotiations and the available scientific knowledge speaks to a problem that the global governance literature calls regime effectiveness. Oran Young (1994) identifies six different kinds of regime effectiveness, but here I am concerned with only one—problem solving: will the climate governance regime have an effect on the climate system, and slow or prevent dangerous climate change in line with Article 2 of the UNFCCC? In contrast to the existing literature (Underdal 1992; Mitchell 2006; Young 2011), my observations concern the stage of regime development rather than a post hoc assessment of international regime effects. While the climate governance regime is in the making, are its creators considering scientific information that is relevant for the effectiveness of future climate governance instruments? In essence, this is a question about a potential mismatch between scientific problem concepts and political solution concepts in the cognitive patterns of climate change negotiators.

The structural disconnect between climate science and the rest of the belief systems of study participants demonstrates that the climate change negotiations prior to Paris were all about getting a deal—a political solution—rather than addressing the climate change problem in the way we have come to understand it scientifically. The politically feasible has trumped the necessary. The mere fact that the Paris negotiations resulted in an agreement has been hailed as a success, and has silenced some of the pessimistic voices concerning multilateral climate governance. Surpassing the low expectations of most negotiators, one could almost hear a collective sigh of diplomatic relief after COP 21. Still, at this point, the international and scientific communities do not know whether the agreement will be effective in the sense that it will slow climate change and prevent climate change–related harm and suffering.

The Goal of Global Climate Governance

Any discussion of regime effectiveness has to start with the question of the ultimate goal of global climate governance. What is the raison d'être of the UNFCCC? What are the parties to the convention trying to accomplish? Although this fundamental question is a necessary starting point for institutional design and norm development, the interview data show that most study participants did not have a clear answer to it. Instead of a shared purpose related to global public goods—climate stability, the avoidance of harm, justice, and so on—study participants focused on things like emission reductions, temperature targets, and energy transitions. When asked what emission reductions were good for, they were tongue-tied.

Article 2 of the UNFCCC states that the purpose of the convention is to achieve the "stabilization of greenhouse gas concentrations in the atmosphere at a level that would prevent dangerous anthropogenic interference with the climate system." This technical expression is the closest the convention ever gets to formulating the ultimate purpose of global climate governance. In an attempt to define what it meant to "prevent dangerous interference," the parties agreed in Copenhagen in 2009 that global average temperature rise should be limited to 2°C above preindustrial levels.

Since the temperature goal was formalized in 2010, a number of scientific studies have been published suggesting that staying below 2°C is still technically feasible, and outlining policy scenarios and emission reduction pathways that would achieve this goal (Rogelj, Hare, et al. 2011; Rogelj, McCollum, et al. 2011; Peters et al. 2013). According to the IPCC's Fifth Assessment Report, limiting global warming to 2°C can only be achieved if global carbon emissions are halved by the middle of the century and eventually reduced to zero. Although global CO_2 emissions reached 32.2 billion metric tons in 2013 and have remained flat in 2014, 2015, and 2016, the prospects for a 50 percent reduction over the coming three decades appear rather remote. Further, these studies are based on critical assumptions, mainly the large-scale deployment of carbon capture and sequestration technologies that can draw emitted CO_2 out of the atmosphere. Without such negative emissions there is hardly any climate model that can demonstrate a pathway to 2°C, let alone to 1.5°C. According to the most recent modeling efforts, the emission cuts necessary for holding global temperatures to no more than 2°C or 1.5°C above preindustrial levels are very similar, but significant reductions would have to occur before 2030 to achieve the 1.5°C goal. Despite these sobering facts, only a few scholars have raised concerns about the target, whether 2°C or 1.5°C, as it is currently defined (Knopf et al. 2012; Jordan et al. 2013), and suggested that the target might have to be adjusted (Geden and Beck 2014) or replaced with a different set of measures.

Even though the assumptions for achieving the global temperature goal are unrealistic and become more so every year, most UNFCCC negotiators draw hope from the general scientific message that achieving both the 2°C and 1.5°C goals is still a possibility. Kevin Anderson (2015) has strongly criticized scientists for helping to create and maintain these overly optimistic beliefs among policy makers with unrealistic model assumptions and a lack of transparency and clarity in their political communications.

The 2011 Conference of the Parties (COP 17) acknowledged that it might be desirable to pursue a more stringent temperature target of 1.5°C, but the parties did not make a commitment to do so. COP 17 also decided that both the adequacy of the temperature goal toward the objective of the convention and progress toward the goal should be reviewed periodically. The report for the first of these periodic reviews (2013–2015), issued at the end of a Structured Expert Dialogue (SED) in 2015, approved of the use of a temperature goal as a metric of success, but cast doubt on the adequacy of the 2°C goal to prevent significant climate change–related harm (FCCC/SB/2015/INF.1). Referencing a range of significant climate change impacts that were already occurring (at 0.85°C global warming), the SED argued that the 2°C goal should be seen as a "defence line" against significant impacts, but that it was not appropriate to think of a 2°C world as safe from such impacts. A long-term temperature goal of 1.5°C would be preferable because it would significantly reduce climate change risks, albeit at a higher cost in terms of necessary carbon capture and sequestration technologies.

The conclusions of the 2013–2015 review provided important leverage for AOSIS and the Climate Vulnerable Forum, a coalition of forty-three highly vulnerable countries, to push for the 1.5°C global temperature goal in the Paris negotiations. To my astonishment, the 1.5°C goal found its way into the Paris Agreement; the Paris COP also invited the IPCC to prepare a special report on the impacts of and pathways toward achieving this goal.

Given these developments, there is little doubt that the long-term global temperature goal will remain a focal point of future negotiations. But the Paris Agreement also introduced a number of additional global goals that have yet to be defined or measured, most notably, "increasing the ability to adapt to the adverse impacts of climate change," "foster[ing] climate resilience and low greenhouse gas emissions development," and "making finance flows consistent with a pathway towards low greenhouse gas emissions and climate resilient development" (Article 2.1 PA). It remains unclear, however, just how progress toward these additional global goals, which do not readily lend themselves to measurement or quantification, will be determined.

But how did study participants themselves think about the goal of global climate governance (in particular, the temperature target) when I interviewed them in 2012? When asked about the goal, only six mentioned the well-being of humanity. The responses differed significantly in their framing and specificity: "to minimize damage and suffering," to maintain "a biophysical system supportive of nine billion people in 2050," "to prevent catastrophic impacts," "to ensure that we can all survive," and "the number that prevents war." Two participants mentioned the convention objective of avoiding dangerous climate change; many more referred to the 2°C target as the final goal of the negotiations. Other responses included ideas like "the decarbonization of the economy" or simply "global GHG emission reductions." Participants seemed puzzled when I asked why decarbonization or 2°C were desirable goals. The temperature target or a vaguely conceptualized system transformation had become goals in themselves, without clear linkages to social, economic, or environmental benefits.

In short, the majority of study participants did not think about the goal of global climate governance in terms of creating public or social goods, such as the prevention of harm and suffering, the avoidance of economic damage, or the stability and well-being of societies. In their minds, the pursuit of technical steps toward these ultimate goals had replaced the goals themselves. Hulme (in Knopf et al. 2012, 124) confirms that this is true for most climate change negotiators and argues that the global temperature target is "socially regressive" by "confusing ends with means."

The cognitive pattern of most participants—a focus on the means to an end rather than the end itself—is not trivial and speaks to the major challenges governments face in communicating climate change policy. Because a global average temperature target does not mean the same to politicians and voters as saving lives and avoiding damage, it does not have the same motivational power. The technical nature of the goal makes it hard for people to imagine it (Hulme in Knopf et al. 2012, 123). Further, the focus on what Hulme calls an "output variable" rather than the various anthropogenic input variables that create global warming adds a significant amount of ambiguity. It is uncertain how—at what level of GHG concentrations—a global temperature goal can be reached. What is more, a temperature goal might simply be unachievable because it requires a higher level of understanding and control of the climate system than humankind currently has or is likely to have in the foreseeable future (Hulme in Knopf et al. 2012, 123).

Many study participants argued that scientists determined the 2°C temperature target as the number that would prevent dangerous climate change, and that political actors simply accepted it. In line with this view, the Copenhagen Accord refers to "the scientific view that the increase in global temperature should be below 2°C." Scientists, on the other hand, emphasize that the selection of a certain target is a value judgment that has to be made by politicians (IPCC 2001). The literature traces the goal back to the European Union's political efforts since 1996 to energize and guide the negotiation process (Tol 2007; Randalls 2010). In 2009, 2°C was a number that appeared to be both scientifically and, according to the 2009 Stern Review, economically feasible and that was likely to avoid mass extinction of species. It represented a pragmatic compromise. Regardless of which origin story is most accurate, they all suggest that the temperature goal did not represent a shared understanding of what constitutes dangerous climate change. Taking this argument one step further, Scott Barrett and Astrid Dannenberg (2012) contend that the parties to the Paris Agreement knowingly set a target that was too weak to prevent catastrophe and pledged less than what was needed to reach the target; worse still, the parties were most likely to do less than was necessary to fulfill even these inadequate pledges. The Paris Agreement's inclusion of 1.5°C casts doubt on Barrett and Dannenberg's first contention, but the nationally determined contribution (NDC) process so far has supported their second. Collectively, the pledges made by all parties to the agreement are likely to lead to a global temperature rise of 3°C. And it remains to be seen whether countries will live up to their promises or whether they will commit to more ambitious NDCs over time.

Even more interesting than the sole focus on the global temperature goal is the fact that, in 2012, most study participants were no longer confident that 2°C goal would be reached. They believed that keeping global warming below 2°C, though both desirable and feasible, was politically impossible to achieve.

If these participants had previously conceived of the 2°C mark as a boundary between safe and dangerous levels of warming, they must have concluded that the world was headed toward grave danger, perhaps even catastrophe. But this was simply not the case. Despite believing that the 2°C target would not be met, these study participants neither despaired nor expected catastrophe. They remained confident and hopeful that the climate change problem would be solved. The coexistence of belief in failure (to prevent dangerous climate change) and hope for future success suggests that they did not associate missing the temperature target with impending catastrophe—things would get worse, yes, but they were not yet hopeless. Participants said things like "Some action is better than no action," "Small wins are possible," or "Four degrees is better than six degrees." They did not simply divide outcomes between those of a benign below-2°C future and those of a catastrophic above-2°C future; they considered a spectrum of outcomes, where 2°C was better than 3°C, but 5°C was also better than 8°C. In other words, they did not perceive of the 2°C temperature target as a tipping point between two qualitatively different worlds—a safe one and a dangerous one—but as one of several points on a spectrum between safe and dangerous.

The inclusion of the 1.5°C goal in the Paris Agreement can only be explained in terms of political interests. Given the higher chance of preserving their existence as states, nations, and cultures in a 1.5°C global warming scenario, it is absolutely in the interest of AOSIS and other highly vulnerable states to insist on this goal. Cognitive science suggests that islanders' minds work hard to counter doubts about the achievability of 1.5°C, looking for sources of hope such as an "overshoot and return" option, as long as there is territory left to defend from the rising seas. Their interests might be better served negotiating for real money (loss and damage) rather than highly uncertain global warming goals. But in extreme cases like global warming, people do not trade off sacred values for money. As outlined above, our minds use deontological reasoning when it comes to matters of survival—a principled and highly emotional moral framing of right or wrong rather than consequentialist reasoning in terms of possibility or affordability.

For other countries, it was politically expedient to confirm the desirability of limiting warming to 1.5°C, even if there was no serious intention to achieve this more stringent goal. The symbolism of support for the most vulnerable parties aided the negotiation process, and no individual country could be held responsible for missing a shared global goal. In other words, including 1.5°C in the agreement text had no political costs but large tactical benefits, facilitating the adoption of the Paris Agreement.

One could argue that there is no harm in aiming high and falling a little short, but there are serious drawbacks to establishing an unachievable goal. For one, the psychology of goal achievement suggests that setting unrealistic (rather than ambitious) goals usually leads to resignation—nobody even attempts to achieve the impossible. Further, the parties might have paved the way for a set of future contestations over which countries were responsible for failing to achieve the shared goal. This could easily be linked to contested questions of loss and damage and the much-dreaded idea of compensation.

Dealing with Complexity

Arild Underdal (2010, 386) argues that some environmental problems are hard to solve because they are "long-term policy problems with time lags between policy measures ... and effects," they "are embedded in very complex systems" surrounded by uncertainties, and they "involve global collective goods" not subject to single best-effort solutions. Climate change is one of these long-term problems embedded in complex systems. How did study participants deal with the characteristics of the climate change problem, its pervasiveness, its nonlinear behavior, time lags, cross-scale interactions, cascading failures, and sometimes irreducible scientific uncertainties?

Study participants in all six groups used the term "complexity" when describing climate change, although in a sense quite different from that used by complex-systems scholars. Nevertheless, despite their limited understanding of complex systems characteristics in a scientific sense, they recognized that there were multiple interacting causes and effects of climate change across multiple social scales, and that climate change was a uniquely pervasive problem that affected all kinds of social and economic behavior. Members of groups 2-MH and 3-LH and, to a lesser extent, 1-HH and 6-LL, were particularly aware of socio-ecological system links. They saw rain-fed agriculture in developing countries, fishing, and nature-based tourism not only as key income and GDP sources, but also as the foundations for livelihoods and ultimately happiness. They were much more aware of the link between nature and economic well-being than representatives of developed countries were. Most members of groups 1-HH and 2-MH had a planetary-systems perspective that linked either all of humanity, or humanity and "all life on Earth."

Most study participants also acknowledged the existence of climate change uncertainties. None of them considered this fact an obstacle for negotiations, however, or a reason to delay action on climate change, although some worried that impact uncertainties posed obstacles to mobilizing public support for such action, and one participant argued that such uncertainties made it difficult to create legitimate adaptation budgets and funding requests. But only one participant made a strong case for the precautionary principle based on scientific uncertainty, arguing that our lack of understanding regarding biodiversity and the importance of certain species at risk of extinction was truly disconcerting—we did not even know what we were losing and how important that loss might be for humanity's well-being.

Imperceptibility of climate change was not an issue for most study participants. Many believed that they had observed climate change impacts at home, and only three stated explicitly that they had not. But some participants, especially those in groups 3-LH and 4-HL, recognized that imperceptibility was a serious problem when it came to rallying public support for action on climate change in the developed world. They argued that personally feeling or seeing the effects and destruction climate change could bring in the form of intensifying and more frequent extreme weather events could change people's minds about the problem. This observation points to the cognitive challenge of dealing with abstract scientific information as the sole motivator for political action in large parts of the developed world, where significant climate change impacts have not yet been felt, where changing weather patterns have not yet been explicitly connected to climate change, and where climate change is still only an idea or a concept, but not a reality.

Given the fast-growing scientific interest in thresholds or climate tipping points over the last decade, it is surprising that most study participants did not yet have a good understanding of the concept and its potential

implications for global climate governance. Many did not mention tipping points or mentioned them only when specifically asked, and when
they did, it was usually in the context of discussing climate science, but
rarely in the context of climate change impacts—and never in the context
of the global climate governance goal or policy design. Participants who
had more detailed thoughts about climate tipping points were primarily
concerned about the irreversibility of climate change associated with them.
A small group of participants (members of group 6-LL and environmental
NGOs) believed that, being both dreaded and needed, tipping points might
be useful to mobilize public support. Development NGO and industry representatives had the most sophisticated understanding of tipping points,
but, even in their minds, the phenomenon was linked to ecosystem change
only, lacking clear connections to change in social, political, and economic
systems.

These observations contradict Mark Nuttall's assertion (2012, 97), at
least as far the climate change negotiations are concerned, that "the tipping
point thus becomes tremendously powerful in discursive, rhetorical, and
metaphorical senses." Instead, current science-policy interactions in the
UNFCCC have failed to create a timely and sufficiently deep understanding
of tipping points and other important scientific concepts among climate
change negotiators (Young 2011).

Finally, members of group 1-HH and representatives of development and
environment and market NGOs displayed a sophisticated understanding
of complex systems dynamics, including tipping points, hysteresis, path-
dependence (dependence of systems' outputs on past and present inputs),
and the need for adaptive governance, adjusting goals and policy measures
over time. Time and urgency played a major part in the belief systems of
these individuals, in contrast to those of most other participants, which
suggests an interesting connection between complex-systems thinking and
environmental concern (Lezak and Thibodeau 2016).

Time, the Future, and a Dearth of Imagination

Political decision making on climate change is in many ways an exercise of
the imagination. Decision makers not only have to imagine the phenomenon of climate change itself, which none of us can perceive or observe in
its totality with our senses. They also have to imagine what kinds of effects
climate change will have on future peoples, societies, plant and animal

species, ecosystems, even the entire planet and how changes in our climate, our sociopolitical systems, and our technologies will interact with one another over time. The UNFCCC negotiators constantly have to generate clear mental pictures of what could at some point in the future become material or social realities. It is those imagined future realities that can and should motivate decisions today (in addition to other motivators related to the past and present political conditions). They guide decision making in certain directions, either to prevent undesirable possibilities or to realize desirable ones.

A key component of this kind of future thinking is time. Negotiators need to have a good understanding of the timelines of climate change. How could climate change unfold over time? When are certain climate change–related events likely to happen? Which are likely to happen sooner, in this decade, and which later, in this century? Wrapped up in thoughts about time are beliefs about the effects of decisions, policies, and actions. How fast can a cap-and-trade system affect the concentration of CO_2 in the atmosphere, and when would an effect on global average temperatures be felt? What about the timelines for implementing renewable energy policies and the effects of their implementation on climate change? When it comes to climate change's multifaceted links between present decision making and future conditions for life on Earth, there is no end to the things that need to be imagined. Below I explore how study participants imagined the future and dealt with intertemporal choice.

Time   The nature of climate change draws our attention to a set of specific temporal features of both the climate change problem and the political process that seeks to address it. Climate change–related policy decisions require the ability to understand how the future is linked to the present. This involves grappling with different time scales, timetables for action, time lags, and systemic inertia, peaking (when emissions will reach their maximum levels), and the concept of time discounting (Schelling 2000; Weitzman 2009; Gollier and Weitzman 2010; Jacobs and Matthews 2012). For example, decisions that are taken in the UNFCCC negotiation process tend to work on time scales of a few or at most ten to fifteen years. The climate system, on the other hand, works on much longer time scales, centuries or even millennia. Can short-term politics address a long-term problem like climate change?

Study participants' thoughts on time were vague and inconsistent; there was no clear cross-group cognitive pattern. Many study participants, mainly in groups 3-LH and 4-HL, made no reference to time at all. Members of group 1-HH used a small set of concepts about the future, which they usually saw as no later than 2030 or 2050, and they converged on a desire to reach a global emissions peak between 2015 and 2020. Industry representatives associated with group 4-HL shared an unusually long time horizon, demonstrated by phrases like "100 to 200 more years of fossil fuel consumption," "commodity price increases in 20 to 50 years," and "planning as far out into the future as one can think." All seven members of group 5-ML experienced a sense of urgency and emphasized the importance of long-term thinking in developing solutions. That they expressed their perceptions of urgency with comments like "acute," "2030 is practically tomorrow," and "race against time" was particularly surprising given that most of these individuals lived in highly developed countries, far from the vulnerable places of the world that had already experienced impacts. Equally interesting was their ability to think about the long term, for example, "impacts on generations to come," "the future beyond grandchildren," "build for the future—50 to 100 years." But even these individuals, who had a real sense of urgency, were unable to present a clear action timeline to address climate change. They did not know what would be "too late" or "in time" for global climate governance.

For most study participants, beliefs about time and intertemporal choice came down to the idea that sooner was better than later and later better than never.

After the Durban negotiations in 2011, the year 2020 became a focal point for many study participants. The reason for this shift was not scientific, but political: the Durban Platform for Enhanced Action stipulated that a new international agreement should be in effect by 2020. Anything sooner than that was considered politically impossible. The biophysical changes expected to occur between 2012 and 2020, or between 2020 and a future point in time when implementing the new agreement would actually affect the climate system, did not figure into the ADP negotiators' thinking. In other words, they did not connect decision-making timelines with governance outcomes. Instead, political feasibility considerations—how long it would take to draft an agreement and to get it ratified—were the main determinants of the Durban decisions.

The study participants' views on this timeline were mixed: some considered it progress, whereas others believed that an agreement that did not go into force before 2020 would be too late because global emissions should have peaked by 2017. Regardless of such beliefs, institutional inertia put the UNFCCC negotiation process on a timeline to 2020. Surprising many observers, this timeline shifted forward when the Paris Agreement entered into force on November 4, 2016, only one year after its adoption at COP 21. The double-requirement of 55 ratifying countries covering 55 percent of global GHG emissions had been fulfilled much faster than anticipated.

Imagination   Beyond these general difficulties in dealing with the temporal aspects of global climate governance, study participants find it particularly challenging to imagine the distant future and how it might be affected by climate change. Imagination is both a cognitive and a sociopolitical process—it takes place in our minds, but also depends to a large extent on our rich and multifaceted interactions with other human beings, institutions, and the natural environment. In the context of climate change, climate science ought to play a significant role in shaping our thoughts, beliefs, and mental images of the future, but, based on the responses and comments of study participants, it does not appear to. When it came to thinking about time, what mattered most to many participants was the present—the demands, concerns, and statements of their own and other delegations that day, that month, or at the upcoming COP. Their dearth of imagination about the distant future and the multiple ways in which climate change could shape that future suggests that, among climate change negotiators at least, the motivational power of scientific insights concerning possible and likely futures is limited in important ways.

Climate change poses a challenge to our imaginations in many ways: much of it cannot be experienced in one place at one time, most of its consequences are expected to take place in the future, and recognizing climate change in the present requires a certain kind of scientific understanding that is hard to achieve. Imagination is not only necessary to make present and future climate change impacts real and perceivable, it is also necessary to identify actions on climate change and solutions to the problems it presents (e.g., policy measures, alternative behaviors, or alternative forms of economic or social organization) that currently do not exist.

One interview question targeted the issue of imagination and climatically changed futures. It challenged study participants to imagine a worst-case scenario of a 2080 world where global climate governance efforts had failed and continued GHG emissions had led to the worst climate change impacts they could imagine. What would their hometowns, their regions, their countries look like in 2080? The question generated a surprisingly consistent response pattern across all participant groups. With very few exceptions, participants said that they had not thought about a worst-case scenario and found it challenging to come up with one on the spot. When asked to imagine such a scenario nevertheless, they rejected the idea of a worst case because "by then [2080], we will certainly have solved the problem." And, finally, when pressed to accept the hypothetical possibility of a worst-case world, many attempted to evade the cognitive challenge by arguing that, between 2012 and 2080, new technological solutions would emerge that neither they nor anyone else could even think of at that time.

Study participants who tried to imagine a worst-case 2080 scenario fell into two groups. Those who did so for the first time during the interview tended to offer linear extensions of current or known climate change impacts: more floods, more droughts, more hunger and poverty, more extreme weather events, and occasionally more resource conflicts. And those (a much smaller group) who were not surprised by this question and had considered a worst-case scenario before presented extremely negative images of a dark and scary future world with fewer states, more violent conflicts, resource scarcity, and a great deal less happiness. Both groups often referred to movies like the 2004 film *The Day after Tomorrow* to help me visualize what they imagined.

The cognitive patterns of study participants suggest that many climate change negotiators either see no reason or lack the ability to imagine the distant future and that their dominant cognitive pattern is the denial of even the possibility of climate governance failure (Milkoreit 2015). From an observer's perspective, this is surprising given the repeated failure of the UNFCCC negotiations to produce tangible results over a period of two decades and the fact that many study participants had been part of those negotiations—which they described as both frustrating and slow—for several years. What could explain this pattern? Given the strong negative emotions associated with the dystopias described by some participants, it is possible that the avoidance of images of the distant future is a cognitive self-protection mechanism, one that might be productive in the sense that it allowed them to work on climate change without becoming depressed or despairing. On the other hand, the lack of imagination and the rejection of the possibility of failure imply that negotiators might never be fully cognizant of what is at stake in the negotiations and what they are collectively putting at risk. The ease with which thoughts about potential future damages are suppressed might significantly limit the motivation of negotiators to develop effective climate governance institutions.

Research in social psychology describes these cognitive patterns as distancing—an active mental process that represents very hard problems like climate change as something distant from the individual or the in-group (Norgaard 2006a, 2006b) or that dismisses such problems out of hand (Wagner 2012). Although the literature tends to focus on risk perceptions regarding present rather than future problems (Spence et al. 2011), it has identified four interacting dimensions of distancing: social (i.e., the problem is perceived to concern other groups), geographical (i.e., it is a problem in far-away places), temporal (i.e., it will happen in the future), and based on uncertainty (i.e., it is unclear whether harm will really occur; Liberman and Trope 2008; Spence, Poortinga, and Pidgeon 2012). Since individuals tend to distance themselves from climate change in the present, it is not surprising that distancing effects are even stronger regarding climate change in the long-term future. Events that are not expected to occur in someone's lifetime are naturally less important than events that are. However, in the context of the climate change negotiations, this cognitive pattern is highly counterproductive. Given the major implications of action or inaction on the part of this generation for the effects of climate change on generations to come, negotiators bear responsibility for the distant future (Clark et al. 2016). They may decide not to value the distant future as highly as the present or the near-term future in their decisions, but their inability to even consider distant costs as costs is clearly an impediment to good decision making. Both cognitive patterns—the absence of clear timelines concerning action on climate change and the lack of imagination regarding the distant future—strongly limit both the possible feelings of urgency and the likelihood of arriving at solutions. Thinking about temporal dynamics and imagining the future are thus major challenges for climate change negotiators hoping to make a real difference.

Agency and Hope—Key Ingredients for Changing the World

What can be learned from the cognitive patterns of study participants about the way agency is perceived, exercised, and constrained in the climate change negotiations? Who believes they have agency in climate governance and which actors are believed to have it? This section briefly introduces the concept of agency as it is currently used in international relations theory and psychology, before applying it to the findings presented above. I make the case that theories of agency in international relations would benefit from two conceptual expansions. First, insights from psychology and the interview data suggest that the concepts and emotions associated with the past and the future play an important role for actor's decisions and behavior in the present. Second, hope is a relevant dimension of agency that has not yet been fully explored either in international relations scholarship or in psychology.

Defining Agency

Although commonly defined as the freedom and ability to act based on intention or purpose, agency could refer to certain effects or consequences of action, for example, an observable change in the actor's environment. But agency also lies in the potential to influence one's environment, not merely in the act of doing so.

International relations theorists often present a simplified division of their discipline into two camps. The rationalist camp emphasizes the causal power of the material system structure, reducing agency to an actor's rational response to structural constraints and opportunities. The constructivist camp insists on the causal power of different identities, norms, and ideas. What the structure-agency discussion comes down to is a disagreement about the source of causal power and ultimately social change in the international system: are actors structurally coerced by the given material reality that determines their interests, or can they choose to act based on motivations that are, at least in part, independent of system structures? His efforts to bridge this structure-agency gap with the help of the sociological theory of structuration—the mutual constitution of structure and agency (Giddens 1992)—have led Alexander Wendt (1992) to famously assert that "anarchy is what states make of it." In Wendt's view (1999), actors have the ability to choose how they interact with structure (and with other actors) and consequently change it. However, as Jeffrey Checkel (1998, 340–342) has convincingly argued, many constructivists—and I would contend that group includes Wendt—in relying on the causal power of social structures to reconstitute actors' interests, have ignored the theoretically important question of agency.

Psychologists have developed an individual-centered understanding of agency as a belief in individual or collective self-efficacy—the ability to anticipate and predict future events and to direct one's own behavior in a manner that influences the future in line with one's motivations and goals (Bandura 1989). The emphasis on self-generated action rather than reaction, in other words, the causal contribution to one's own motivations and actions, is key to this perspective. Agency begins in the mind. But agency is ultimately the result of multiple interacting factors, including cognition and affect, other personal factors, resource availability, and environmental events. Albert Bandura (2001, 1) presents a social cognitive theory of agency according to which "agency embodies the endowments, belief systems, self-regulatory capabilities and distributed structures and functions through which personal influence is exercised." He emphasizes the temporal dimensions of agency and its extension into the future through forethought, intentionality, self-reflectiveness about one's capabilities, and the meaning and purpose of one's life pursuits. Adopting a form of structuration theory, Bandura (2001, 1) argues that "people are producers as well as products of social systems."

Jointly, the international relations and psychological perspectives provide a good analytical foundation for addressing the role of agency in the global climate change negotiations. The international relations lens focuses on the interaction between an actor—often assumed to be a unitary state—and the structure in which the actor is embedded (i.e., the international system). More often than not, scholars of international politics argue that the structure determines or at least constrains agency in important ways. But though they assume that actors have the ability to interact with and change the given structure, they do not often explore this ability empirically. The psychological lens focuses on the individual—usually a person rather than a collective entity—and the origins of agency in the cognitive-affective system that interacts with its structural environment and ultimately seeks to influence that environment through forethought, goal setting, and intentional, purpose-driven behavior.

Contrasting these two disciplinary perspectives reveals an important conceptual gap between individual and collective agency, for example, between the statement of a single diplomat and the behavior of a state. Not only do international relations scholars attribute agency to groups, but so do ordinary people in their everyday language and especially diplomats in UN negotiations. International relations scholars have hardly addressed collective agency, however, apart from legal representation theorists, and even they use the fiction of the state as an entity separate from those who act on behalf of the state, which supposedly has a will and consequently agency. Although individual agency is based on a person's individual thought processes, collective agency has no such basis. So how can we attribute agency to collective entities like Germany, Exxon Mobile, Occupy Wall Street, or Greenpeace?

As outlined in chapter 3, the bridge between the individual and the group exists in the minds of individuals, both group members and outsiders, who have mental representations of the group, its characteristics, membership criteria, purposes, and so on. These mental representations are shared among many individuals through communication and interaction with the social and natural world, which means that different group members and nonmembers have the same cognitive patterns with regard to the group. Using these mental representations, individuals enact the group through their decisions and behaviors. Since the actions of groups result from the actions of individuals, who think about the group, groups attain agency through individual cognition. At the same time, individual cognition is never isolated from its social and physical environment.

The notion of collective agency based on individual cognitive processes strengthens a number of insights presented earlier in this chapter, for example, the ability of individual diplomats to think about climate change impacts in their home country or even elsewhere in the world, and to experience emotions in response to collective loss or threats from such impacts, whether deaths due to extreme weather events or economic damage or threats to place identity. The cognitive explanation of collective agency will also be a useful guide for the following sections on agency in the climate change negotiations.

Agency and National Interests

Intentionality, purpose, and motivation are key concepts of agency. International relations scholars think of states' motivations as national interests. Rationalist scholars assume that national interests are determined by the material system structure, which presents actors with constraints and opportunities. States survey their environment and make decisions based on a desire to maximize their own net gain, pursuing the path of action with the largest positive difference between expected costs and benefits. Constructivists argue that the externally given system structure does not predetermine national interests. Actors have to interpret and attribute meaning to the system structure, which they do from the vantage point of a certain identity, associated norms, and beliefs. Thus normative and reputational concerns can shape national interests as much as the desire to maximize material gain and power.

Thus, too, both rationalist arguments about costs and benefits and constructivist arguments about identity and norms powerfully shape the definition of an actor's interests, where interests = expected net benefits × identity × norms. It is impossible to define an actor's interests without first identifying the group that actor belongs to and seeks to protect. Group affiliation determines the type and temporal proximity of a threat the actor perceives (i.e., potential costs and benefits of action and inaction). These threat perceptions in turn determine the ethical framework activated in the actor's mind (deontological or consequentialist), which triggers associated norms and norm interpretations. When combining these conceptual categories across theoretical boundaries, it becomes clear why group 3-LH pressed for immediate action, group 4-HL was reluctant to commit, and group 1-HH found itself in the pragmatic middle ground between these different definitions of national interests. Depending on the constellation of identity group and concerns about cost, different levels of cognitive disturbance lead to very different moral sentiments, ranging from cold, seemingly unemotional utilitarian thinking to emotionally charged normative thinking.

Further, the data suggest that it is not always national interests that matter in international politics. Some study participants combined, switched between, or solely relied on alternative group identities and their respective interests, whether these interests were humanity sharing one planetary resource or vulnerable people in need of protection from climate change impacts induced by other humans, who had the resources to help.

Agency and Time
The cognitive lens also reveals that international relations scholars have ignored an important dimension of agency, which psychologists have already begun to grapple with: time. Theories of agency will have to be expanded to account for the relevance of the past (memories) and the future (anticipation) for decisions and behavior of political actors in the present. Emotions play an important role when theorizing about time and agency because mental representations about the past and the future can be associated with different types of emotions and their varying levels of intensity. These emotions can impact the salience of certain beliefs for decisions taken in the present.

The data presented here crystallize the importance of two types of time-related cognitive elements for belief systems about climate change. First, memories, in particular memories of places that are changing due to climate change, and the emotions associated with those places, heavily influence both identity conceptions and threat perceptions. Second, anticipated future climate change, in particular a lack of imagination regarding the distant future, affects motivations to act in the present. Both of these cognitive elements, memories and anticipations, can enter our thoughts or decision-making process, but they do so in different ways.

Memories tend to be about things, people, and places that we have seen, experienced, and felt. They are associated with images, smells, textures, and other sensory experiences that enable us to make them vivid in our minds. In the case of climate change, memories are often about things and places that certain people might be losing—positive elements of their past that climate change threatens to remove from their current physical reality. Although the memory of those places is certain and cannot be taken away by climate change, the potential loss of the real places linked to the memories affects the way those affected think about the past—it might become more precious. Memories are also linked to the future because the loss of familiar places makes it impossible to pass a certain place and its experience on to one's children. This loss of intergenerational stability is another form of ontological insecurity.

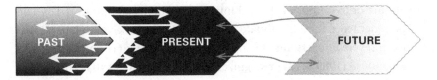

Figure 5.4
Role of time in decision making.

In contrast to the certainty and colorful, emotional vividness that our minds can activate with memories of the past, thoughts about the future do not benefit from the cognitive intensity that accompanies experience. Instead, anticipation of the future depends to a large degree on our ability to imagine and visualize things that have not yet come to pass (see figure 5.4).

Social cognitive theory addresses the concept of time in the form of forethought and goal setting, which are symbolic cognitive activities. Bandura (2006, 164) emphasizes that these symbolizing capabilities enable us to transcend the present, shape our life circumstances and even "override environmental influences." The future cannot be causal for behavior in the present because it has not happened yet, but its cognitive representation in the present can motivate decision making and behavior (Bandura 1989, 1179). The capacity to extrapolate future consequences from known facts enables us to take corrective action to avoid future harm, which presumably increases the prospects for human survival (Bandura 1989, 1181).

Not only concepts but also emotions connected with mental representations of the future matter. We savor anticipated good events and dread bad ones ahead of their occurrence (Loewenstein 1992). Dreading serious climate change impacts—to the extent that such impacts can be anticipated (i.e., imagined) and emotionally experienced—should therefore motivate us today to make choices that prevent future harm.

The findings presented earlier suggest that the easiest way for our minds to deal with the challenge of imagination is to use the experience of the past and extend it into the future, even if there is knowledge that the future will be different from the past. Indeed, the relative weakness of our imaginative abilities regarding the future and the weakness or even absence of associated emotions might prevent us from using our cognitive survival power. This weakness affects our ability as individuals or in groups to value the future or to estimate expected future costs (Berns and Atran 2012). The unprecedented nature of climate change and the perceived temporal distance of severe impacts might simply exceed our current cognitive capabilities. Bandura (1989, 1181) recognizes this when he states that "the power of anticipative control must be enhanced by developing better methods for forecasting distal consequences and stronger social mechanisms for bringing projected consequences to bear on current behavior to keep us off self-destructive courses."

In a nutshell, these theoretical and empirical insights imply that, even though the past might have a fairly strong influence on the deliberations of climate change negotiators today, the more distant future might not even enter into their cost-benefit calculations or normative considerations. In addition to grappling with agents' relationship to structure, agency theory within international relations theory should include a cognitive dimension that can link time (the past and the future), place, and emotion to decision making.

Agency and Hope

The interview data combined with observations of the UNFCCC process reveal a number of interesting things about the role of hope in the climate change negotiations. Most study participants were fairly pessimistic regarding the effectiveness and speed of the UNFCCC process and the chances of achieving the 2°C goal, but they remained surprisingly optimistic about humanity's collective ability to solve the climate change problem. Participant negotiators apparently recovered fairly quickly from feelings of hopelessness: some had experienced such feelings at the 2009 COP in Copenhagen or the 2011 COP in Durban, but not for long. And the general tone of diplomatic discourse on the United States remains hopeful despite the minimal contributions that country has made to the climate governance regime over the last fifteen years.

Charles Snyder (2002, 249) defines hope as "the perceived capability to derive pathways to desired goals, and motivate oneself via agency thinking to those pathways." Using insights into the pursuit of goals, psychologists have established a strong theoretical link between hope and agency. Victoria McGeer (2004, 100) argues that hope is "an essential and distinctive feature" and even a "unifying and grounding force of human agency." As Snyder's definition indicates, hope is a motivational state that combines the future-oriented setting of a goal—"the cognitive component that anchors hope theory" (Snyder 2002, 250)—with the identification of pathways (actions) toward that goal. Problems are seen as obstructions on the paths to people's goals; they decrease hope and consequently weaken perceptions of agency. Not many political scientists have explored this topic, but those who have tend to address questions of collective agency and social mobilization. Sasha Courville and Nicola Piper (2004) suggest that hope can have empowering effects and that NGOs can use hope to mobilize political support for social change. Conflict researchers have explored the interaction between hope and fear, finding that memory-based fear tends to overpower anticipation-based hope in conflict-ridden societies (Bar-Tal 2001).

But even with this theoretical guidance, it remains unclear how hope, agency, and goal setting interact with one another. Causal arrows between all three elements are possible (see figure 5.5). First, hope can affect agency, both positively and negatively (i.e., increasing hope increases one's sense of agency, whereas losing hope results in a loss of agency.) Second, agency can affect hope: a strengthened sense of agency can increase hope, and a weakened sense of agency can diminish hope.[2] Third, both agency and

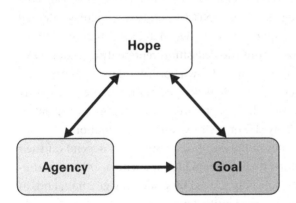

Figure 5.5
Relationship between agency, hope, and goal.

2. Viktor Frankl's account (1959) of his concentration camp experience in *Man's Search for Meaning* offers a celebrated example to the contrary. As an Auschwitz and Dachau inmate, Frankl had absolutely no control over his own life, or even daily activities; yet, despite this lack of agency, he maintained a sense of hope and found meaning in existence.

hope interact with the setting of goals: the type of goal determines the available pathways toward it, whereas one's own skills and abilities determine what kinds of actions one is able to take, making certain goals more or less achievable.

The causal relationships between hope and agency and between hope and goal setting are bidirectional (indicated by double-headed arrows in figure 5.5): they affect each other depending on their current state and strength, whereas the causal relationship between agency and goal setting is unidirectional (indicated by a single-headed arrow): a sense of agency influences the selection and pursuit of a certain goal, but not the other way around. The reverse relationship between goal setting and agency is constitutive rather than causal (hence not indicated by an arrow in the figure): the form and sources of agency depend on the goal that is set.

The links between agency, hope, and goal setting are asymmetrical in the sense that different factors can influence each corner of the triangle. Agency-related beliefs can be influenced by scientific knowledge, perceptions of resources and power, causal beliefs, available technologies, and uncertainty; concepts of efficacy are of central importance. In a collective-action dilemma like the climate change negotiations, the individual negotiator's assessment of collective efficacy (Bandura 2006, 165–166)—the joint ability of all engaged in the UNFCCC process to achieve desired outcomes—is at least as important as the negotiator's assessment of his or her individual efficacy or that of the delegation representing a country or NGO. Because the effectiveness of their own contributions to climate governance always depends on what others do, negotiation parties constantly have to assess the likelihood that other parties will support a cooperative solution, as well as the likely size of their respective contributions.

Individuals' mental states of hope or hopelessness can depend on their beliefs about human nature, their personal life experiences, or their spirituality. The setting of goals is heavily shaped by values, imagination, and, in the case of climate change, also scientific knowledge.

Analysis of their cognitive-affective maps provides some clues regarding the sources of agency and hope that study participants perceived in 2012, although this might have changed significantly with the adoption of the Paris Agreement. As already mentioned, many participants no longer believed in 2012 that the 2°C temperature goal was still achievable. Given their doubts concerning an effective political solution and their concerns about the sluggish pace of socioeconomic change, they were not able to conceive of reasonable pathways toward the 2°C goal.

When asked what made them hopeful despite these doubts, individual participants pointed to a variety of factors within and without the negotiation process. These included past successes in creating an institutional architecture (e.g., the Green Climate Fund), recent developments and the potential for change in the private sector (i.e., others' agency without personal involvement or control), community activities based on a shared desire to create a better future (i.e., collective agency at the local level with personal involvement), and the belief that a better future was possible, however unlikely it might seem at the time. Participants also identified a set of hope- and agency-diminishing obstacles to a negotiated agreement. The most prominent among them was the negotiation position of the United States, which study participants believed to be constrained by vested interests and a lack of public support for action on climate change at the domestic level.

Figure 5.6 diagrams the interactions of key concepts for two important phenomena in the beliefs of participant negotiators. First, the US position had a negative influence on others' perceptions of collective agency because it weakened collective efficacy beliefs. The general awareness of both the importance of the United States for global mitigation and its current domestic political handcuffs (combined with other countries' reluctance to "go first") significantly reduced available action options and consequently imagined pathways toward the goal—the 2°C target. This resulted in

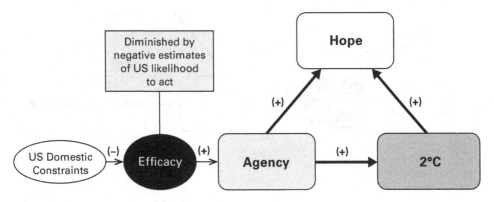

Figure 5.6
Relationship between agency, hope, and goal for the 2°C target.

pessimism regarding the temperature goal. Given his position that climate change is a hoax and his stated intention to "cancel" the Paris Agreement, the election of Donald Trump as the next President of the United States is likely to have significant, negative effects on negotiators' perceptions of collective agency in the UNFCCC in the future. The goal itself was beset by a number of cognitive challenges, including the absence of a clear understanding whether and how the 2°C target could be reached. The negative interaction between agency- and goal-related thought patterns resulted in hopelessness.

Second, in the face of this intractable problem—no study participant in the UNFCCC process had a sense of agency with regard to domestic US politics—individual participants maintained a sense of hope by shifting their definition of what they were trying to achieve. Instead of pursuing the temperature goal or seeking to avoid dangerous climate change, participant negotiators focused on a goal they could achieve with the tools under their diplomatic control: a political agreement (see figure 5.7). Although they did not see a clear path toward keeping global warming below 2°C, they knew how to use the existing political processes and instruments of diplomacy to arrive at a negotiated agreement. The cognitive search for agency and the associated positive experience of hope help explain the mental

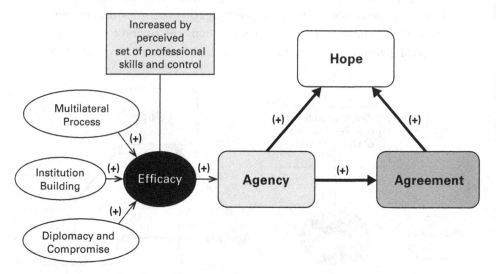

Figure 5.7
Relationship between agency, hope, and goals for a political agreement.

separation between the political feasibility of a multilateral agreement and the scientifically prescribed necessities of action (targets, ambitious initiatives, timelines). The side effect of the hope-agency interaction is a purely political definition of success that ignores regime effectiveness, but facilitates an unreasonable sense of optimism about the UNFCCC process. Kevin Anderson and Alice Bowls (2012) suggest that scientists contribute to such "unsubstantiated hope" and "neglect for serious constraints for achieving goals." However detrimental that cognitive pattern is for regime effectiveness, it certainly facilitated success in Paris.

Summary

No single theory of international relations can explain the way UNFCCC negotiators cognitively deal with and ultimately make decisions on climate change and multilateral cooperation. Even the combination of all major international relations theories does not fully capture the relevant thought processes driving political behavior within the UNFCCC process. But my study's cognitive-affective analysis demonstrates that multiple theories contribute important pieces of the puzzle. More important, the integration rather than separation of these theories creates a deeper understanding of the dynamics of global climate change negotiations. Rational-choice and constructivist cognitive patterns strongly interact with each other and create belief patterns and ultimately negotiation behavior that cannot be fully explained using only one of these theoretical lenses.

The most impressive example of this theoretical interaction is the cognitive relationship between threat perceptions, identity groups, and norms of justice—what I have called the cognitive triangle. Although a simple construct, the triangle accounts for different and complex types of political belief systems that inform negotiation positions. Emotions triggered by perceived threats to negotiators' in-groups play a fundamental role in determining their negotiation stances, affecting in particular the nature of their normative decision framework (either deontological or consequentialist).

The differentiation of threat categories and their relationship to a negotiator's identity brought into focus the importance of place as an identity dimension that is often neglected in international relations scholarship. The nexus between place identity and the expected loss or change of places as a consequence of climate change further illuminates the importance of emotions in the climate change negotiations. Study participants who experienced significant threats to their place identity favored multilateral cooperation with an intensity and urgency that was not matched in the belief systems of those who experienced no such threats. This was true for participants whether they lived in the developed or developing world.

Norms of justice played a major role in the belief systems of all study participants, but did so in very different ways. Consequentialist norms merged with rational-choice frameworks into "cold" negotiation positions centered on national interests and concerns about the costs of action to mitigate climate change. Deontological norms created "hot" negotiation positions informed by a desire to protect people from the threats of climate change impacts.

More generally, participant negotiators presented a much broader spectrum of norms that were significantly less contentious than those of the major normative debates in the UNFCCC negotiations. Instead of equity and static definitions of historical responsibility, the participants' dominant norms focused on broadening the group of parties that should contribute to solving the climate change problem by including all major emitters, the emerging economies, or all countries, and emphasized notions of international solidarity, and a human responsibility to help other human beings in need.

The data also provided some fresh input into the debates on the importance of domestic politics for international affairs, and whether an engaged hegemon is necessary in environmental regime building. The study participants' focus on US domestic politics as the most important obstacle to an international agreement suggests that congressional resistance to any kind of binding international agreement motivated the development of the new hybrid model of global climate governance. Indirectly and certainly inadvertently, an unmovable Congress might have helped create the Paris Agreement.

With few exceptions, the role of science was weaker than expected in the belief systems of study participants. The link between climate science and the ultimate goal of the climate change negotiations was particularly complicated. The 2°C temperature target had largely replaced definitions of the goal or purpose of the climate governance regime as a public good. At the same time, many individuals across all participant groups were abandoning this target, with implications for the political process. Participant negotiators, lacking a sense of agency with respect to the temperature target, resorted to a political compromise or agreement as a default measure of regime success, rather than environmental effectiveness.

The cognitive patterns of study participants demonstrated only a weak and unsatisfactory engagement with a number of characteristics of the climate change problem. Because many participants did not understand some features of the problem's complexity, for example, emergence, systemic lags, nonlinearity, threshold behavior, or cascading failures, they rarely explored the implications of these characteristics for the institutions and policies of global climate governance. Surprisingly, they knew very little if anything at all about the importance of climate tipping points.

Participants' thoughts related to time and intertemporal choice were inconsistent, and most lacked imagination regarding the distant future. Both of these cognitive features interacted and inhibited their considering the full future implications of present decisions about climate governance. The absence of a cognitive link between present action or inaction and future consequences limited their sense of responsibility for the future. Many struggled to comprehend the extension of this generation's causal powers into the distant future, and the responsibility that attended this new anthropogenic power. Without a well-developed public discourse or scholarship on ethical frameworks for such long-term, intergenerational problems, negotiators' minds lack the tools to deal with these challenging features of climate change.

However, study participants were generally confident that the climate change problem could be solved with existing knowledge and governance tools, that uncertainty was no impediment, that the problem's complexity was neither overwhelming nor paralyzing, and that the imperceptibility of climate change was only a problem at the domestic level, where public support needed to be mobilized.

Some of these insights hold important clues to the role of agency in climate governance. The participants' perceptions of a single actor—the United States—severely constrained their perceptions of the collective agency of the international community. The combined assessment of the United States as both powerful and unlikely to cooperate diminished participants' sense of multilateral agency. Many individuals based their assessments of the likelihood of international cooperation on their perceptions of domestic political constraints in the United States. They also linked these assessments to the United States' strategic interests in the international system. Observing an ongoing power transition, many believed that the United States sought to protect the status quo of economic power distribution in the world, trying to prevent emerging powers from gaining competitive advantage through climate change policy. Such pessimism about the UNFCCC process often resulted in a turn to processes outside the United Nations, which offered greater reasons for optimism: other multilateral activities with fewer parties, civil society activities, the private sector, technological developments, including a price drop for renewable energy technologies, or community-based initiatives, although no participant assessed these activities in terms of their effectiveness.

In the years following my empirical data gathering (2013–2015), the United States became a much more proactive player in the climate change negotiations. The Obama administration's efforts to build a strong relationship with China and to jointly push for credible domestic climate change policies undoubtedly had a very positive effect on collective agency beliefs among negotiators, and hence facilitated the Paris Agreement. The election of Donald Trump as the next President of the United States is likely going to have the opposite effect, hampering the implementation of the Paris Agreement.

Further, participant negotiators identified parts of the solution space that existed outside of the abilities and reach of the negotiation process. Recognizing that some of the necessary changes were beyond their control constrained participants' sense of collective agency with regard to the UNFCCC community. Examples included the idea that solving climate change required a new development paradigm, or the need to create support for climate policies among domestic constituencies.

Time is another important aspect of agency-related beliefs that shape decision making in the climate change negotiations. Given the cognitive ease of evoking past memories and reexperiencing associated emotions, but also the cognitive difficulty of imagining the distant future and experiencing emotions associated with that future, memory played a much stronger role than anticipation in the participants' political decision making. Because of the special temporal properties of climate change, this differential impact of the past and the future on present decision making presents a unique and significant challenge to global climate governance.

Finally, a closer look at agency and hope has revealed interesting cognitive patterns regarding the definition of a goal for global climate governance and related measures of success. If a goal appears unachievable, as the 2°C target did in 2012, people lose their sense of agency and hope. Since hopelessness is a negative mental state that people seek to avoid, they turn to alternative goals that appear more achievable. In the case of climate change, participant negotiators replaced the temperature goal with the diplomatic goal of reaching a political agreement. The latter gave them a stronger sense of agency, and agreements like the Copenhagen Accord, the Durban Platform, or the Doha Gateway provided small boosts to their morale. Although this has negative implications for regime effectiveness, it ultimately gave rise to the Paris Agreement.

6 Six Belief Systems—More Alike Than Not?

Rarely concerned with the personal convictions of state representatives, analysts of international politics consider what political actors say and do to be much more relevant than what they think. I beg to differ. In the private belief systems of actors in global climate change politics, which may or may not overlap with their public statements and diplomatic behavior, we may find important clues to the nature of the climate change problem we face, to the reasons for its stubborn resistance to negotiated solutions observed over two decades, and to the pathways out of deep political contestation.

This chapter presents the results of a Q study I conducted in 2012 with twenty-eight participants engaged in the climate change negotiations to explore whether and to what extent they shared any or multiple belief systems with respect to climate change and global climate governance. Although my analysis of their cognitive-affective maps explored individual concepts, their emotional valences, and connections in individual minds, the macro perspective of the Q method revealed whole belief systems—sets of coherently linked ideas that could be found in the minds of multiple participants. Focusing on collective cognition, I used this method to complement cognitive-affective mapping by identifying belief systems that were shared by individual participants across all participant groups. Keep in mind that the belief systems I identified are probably not comprehensive; although they capture a large and interesting set of viewpoints, they likely missed other perspectives that are relevant in the negotiation process. But even though the Q study could say nothing about the relative importance of these belief systems or their distribution in the larger community of climate change diplomats and observer organizations, it offered insights into the kinds and content of existing viewpoints, as well as the connections, overlap, and tensions between them.

General patterns in the Q data set are reviewed in the first section of this chapter; the results of the factor analysis—the heart of the Q method—are presented in the second, which describes in detail six different belief systems (factors). When the results permit, I explore connections to both the CAM analysis in chapter 5 and the factor interpretation. As in chapter 5, I emphasize three cognitive elements in each factor: threat, identity, and justice.

General Data Patterns

Statements Generating Broad Agreement among Study Participants

As outlined in the methodological introduction in chapter 4, the Q method uses a set of opinion statements about the topic in question (the Q set) to elicit responses from research subjects concerning the validity and importance of these statements. In the Q study I conducted with twenty-nine participants engaged in the climate change negotiations, including myself, some of the opinion statements received very similar, extreme rankings from a large number of study participants, indicating broad agreement regarding their relevance among a diverse set of negotiators. Ten statements (out of a total of sixty-five) received consistently high rankings (+4 to +5 or –4 to –5) by more than a third (minimum ten) of the twenty-nine Q study participants (see table 6.1). A positive ranking indicates agreement with the statement; a negative ranking indicates disagreement. High rankings (here, +4 or +5) can imply a high level of agreement or an assessment of the importance of the statement for the participant's belief system.

Three statements stand out, with more than half of the Q study participants ranking them similarly high or low. Statements 1 and 2 concern the very foundation of the climate change negotiations: the reality of climate change, its anthropogenic causes, and the trust in science as the most important source of knowledge about the problem. What is surprising about the scores for these two statements is not their similarity or the broad agreement with them among study participants (not one participant assigned them a negative score) but that, even after more than twenty years of negotiations, these two statements were still considered to be the most important among sixty-five statements regarding global climate governance. Instead of simply treating these as starting premises widely accepted

Table 6.1
Statements with a high number of similar, extreme scores

Statement	Number of scores				Totals	Minimum or maximum score
	+5	+4	−5	−4		
1. Human-released greenhouse gases are causing significant climate change.	7	10			17 (59%)	0
2. I don't trust what scientists say about climate change.			8	9	17 (59%)	0
4. I do not believe that we will see significant effects of climate change in my lifetime.			4	10	14 (48%)	+2
6. Climate change is mainly an issue for the developing countries.			7	6	13 (45%)	0
15. The climate problem should be left to the markets.			5	6	11 (38%)	+4
18. Politicians need much stronger voter support and pressure from political movements to create meaningful climate policies.	4	6			10 (34%)	−4
21. Neither states nor markets nor civil societies can solve this problem on their own—climate change is a multilevel problem and requires action at all of these different levels.	4	9			13 (45%)	−1
22. A key element in solving the climate problem is the need for fundamental value change within our societies.	9	4			13 (45%)	−5
47. Problems that might arise decades from now are not important to me.			6	9	15 (52%)	−1
58. It is already too late to do anything about climate change.			5	9	14 (48%)	+2

among UNFCCC negotiators and ranking them somewhere in the middle of the positive range of scores, study participants felt the need to state strong support for them.

One explanation for this pattern is the cognitive ease participants might have had in identifying these fundamental beliefs as important among the complex set of arguments one could make about climate change, whereas they might have found it difficult to attribute higher and lower importance to the more complex political arguments. The persistent phenomenon of climate change skepticism and denial might offer another explanation. The reality of the climate change problem and the trustworthiness of scientists are still contested in a number of countries, most prominently, the United States. Although these questions are no longer debated among UNFCCC negotiators, stating their convictions on them still seemed to be important and necessary to my study participants in 2012.

The extremely negative scores for statement 47 expressed participants' concern for and acknowledged the importance they attached to thinking about the future when discussing climate change in the present. However, statement 47 is very general and does not specify how or why this concern about the future matters or should be acted upon. The statement cannot be fully understood in isolation, but requires interpretation in relation to other statements and their scores in a participant's full Q sort. That said, the general disagreement among study participants with statement 4 might offer a first clue. If people expect to experience climate change impacts in their own lifetimes (i.e., over the next thirty to fifty years), and have an interest in avoiding those, problems that arise decades from now would be important to them.

Statements 15, 18, 21, and 22 concern the possible policy and governance approaches to climate change. They acknowledge that climate change is a multilevel problem that cannot be solved with a multilateral treaty alone, and that the national and perhaps even subnational levels are extremely important. "Fundamental value change within our societies" (statement 18) and "voter support" and "pressure from political movements" (statement 22) are possible mechanisms of change that can influence a negotiated outcome only indirectly. Further the call for value change (statement 22) but rejection of free-market policies (statement 15) could indicate broad support for a value set not driven by neoliberal ideas that prioritize growth driven by self-interested producers and consumers.

The process of factor extraction reduces the number of comparable scores to six, that is, each of the six factors is now represented by a factor array with a single score for each statement. With only six scores to compare, the range of scores for a statement (e.g., –3 to –5) is a more interesting measure of agreement among participants than the number of factors where participants assigned a statement the same score. Major agreements among participants across the six factors remain, but statements receiving a wide range of scores disappear from the list of statements generating high levels of agreement (e.g., statements 15, 18, and 22). A number of new areas of belief conversion emerge and are included in the list below. The score range for each statement is indicated in parentheses.

16. It is best to leave the development of climate solutions to regions, cities, and local communities—they have been much more successful than UNFCCC negotiations. (–1 to –2)
33. Economic growth and jobs must take priority over climate concerns. (–2 to –3)
47. Problems that might arise decades from now are not important to me. (–4 to –5)
60. Climate change scares me because I don't know what's going to happen. (0 to –1)
64. I have a hard time imagining the consequences of climate change for my community and my country. (0 to –1)

From an analytic perspective, these high-agreement statements, especially those receiving extreme scores, pose a challenge. Extreme scores are crucial for identifying differences between participants in different factors, and the large degree of similarity of these extreme rankings limits the available information for interpretation and differentiation. Consequently, factor interpretation for this study had to become much more detailed, drawing on differences in lower score ranges and even scores toward the center of the distribution. On the other hand, the existence of large areas of agreement bodes well for the negotiations themselves. Although one cannot reliably extrapolate from the Q study to the larger population of UNFCCC negotiators, my Q study findings suggest that there are significant areas of agreement that various parties rely on. It is particularly important to recognize these similarities between participant group 4-HL and other groups, moderating the concern about the differences between them identified in the CAM analysis.

Statements Generating Wide Disagreement

Statements receiving a wide range of scores (maximum of 10 points) indicate major disagreements among study participants. They are important to understand the key sources of contention between different belief systems. Three of the sixty-five statements received scores at both ends of the spectrum (–5 and +5):

12. Only a small number of countries with significant GHG emissions are important for climate negotiations.
22. A key element in solving the climate problem is the need for fundamental value change within our societies.
24. God has made us stewards of the Earth, giving us the ability and responsibility to keep the planet healthy.

Statement 22 was also among the statements generating a high level of agreement among participants, which suggests that most of them supported the statement, and the opposing belief was an exception. Because most also found the topic of religion (statement 24) to be not very important in the negotiations, the disagreement regarding this statement is not very consequential. One participant commented: "I placed the 'God has made us stewards' question into 'neither/conflicted' because I don't strongly identify with the concept of God on a personal level, but worded differently, I would agree with this general statement of responsibility in a spiritual context." Of real importance is the opinion split regarding statement 12—the number of countries that participants thought were important for the climate change negotiations.

Nine other statements received a wide spread of scores before factorization, hence generated wide disagreement:

3. Anti–climate change policies threaten progress and modernity.
14. The power disparities between countries in climate negotiations make me very upset.
15. The climate problem should be left to the markets.
18. Politicians need much stronger voter support and pressure from political movements to create meaningful climate policies.
32. Taxes—whether globally or domestically—and other policies that constrain private property rights are simply not politically acceptable in our system.
38. Economic growth is the best solution to climate change.

41. I am ashamed that my country is not doing more about climate change.
48. All states have a moral responsibility to contribute to a global climate solution.
50. Future generations are likely to be richer and better off than we are, and better able to deal with climate change.
61. Climate change is not the only issue we have to deal with and other issues are often more urgent.
62. The focus on winning the next election is the biggest obstacle to finding international agreement.

The main themes of these statements include the role of the state versus markets and economic growth in addressing climate change, private property constraints as domestic climate change policy instruments, and voter support and elections as obstacles to an international climate change agreement. How these statements mattered to participants and what types of belief systems they supported will become clearer in the process of factor interpretation.

Factorization reduced the maximum range of scores to seven, which could be observed for three statements:

7. Climate change was caused by rich, industrialized countries, but its impacts will hurt poor countries most—this asymmetry is unfair.
41. I am ashamed that my country is not doing more about climate change.
45. My government is a very constructive player in international affairs.

Interestingly, at this stage of analysis, beliefs about fairness and national identity, expressed in terms of shame and assessments of constructive attitudes of governments, were the main sources of differences in opinion.

Other Statements Receiving Surprising Scores
A number of additional statement scores are noteworthy and shed some light on a multitude of persistent issues in the climate change negotiations.

8. Climate change will result in violence and human deaths.

The highest factor score for this statement was +3 and the average +1.8. Despite increasing reports of thousands of people dying due to extreme weather events and despite the proven link between extreme weather events and climate change, the idea that climate change could pose risks to human lives did not yet play a major role in the beliefs of many Q study participants in 2012.

10. States are the most important players in global climate politics.

That the highest factor score for this statement was +2 and the average factor score a mere 0.5 for this group of individuals tasked with multilateral negotiations was somewhat surprising. Twelve participants assigned this statement a negative score. The highest individual score (+4) was given by only two, one of them an NGO representative. Although more than half of the Q study participants represented state governments that sought to address climate change through international cooperative agreements, they expressed only weak support for the idea that states were the most important actors in climate change politics.

16. It is best to leave the development of climate solutions to regions, cities, and local communities—they have been much more successful than UNFCCC negotiations.

Thus participants seemed to have doubts not only about the role of states, but also about the ability of subnational authorities to address climate change. Statement 16 received only negative factor scores (–1 to –2). But, as one participant pointed out, the negative rankings did not necessarily mean that participants believed subnational levels of climate governance were unimportant; rather, the low rankings reflected the problematic framing of the statement: "Whether we should 'leave' climate change to the municipal governments and others, or if we should concentrate 'all resources' on domestic action. Domestic, municipal, city-level action is … where most of the action is, but I can't agree with 'leave it' or applying 'all resources' to them, because the larger context is important for setting the tone and motivation."

19. The vested interests blocking solutions are too powerful to allow for any meaningful action on climate change.

With an average factor score of +1 and a maximum factor score of +2, this statement received less support from study participants than I had expected. This might indicate either that they did not attribute major importance to the blocking power of vested interests or that they believed the power could be overcome.

26. Limiting average global warming to 2°C will be sufficient to prevent major damage.

The factor scores for this statement ranged from –1 to –3. One participant assigned the highest raw score, +1; eight scored this statement 0. These rankings suggests that, at least in 2012, participants had little confidence that reaching the temperature target of 2°C would achieve the objective of the UNFCCC. On the one hand, this raises doubts about the utility of and general support for the 2°C goal that goes beyond political declarations. On the other hand, it explains why many countries were supportive of including a reference to the more ambitious goal of 1.5°C in the Paris Agreement. Limiting warming to 1.5°C would prevent more damage than 2°C would. At the same time, participants' Q rankings gave very little weight to the following statement:

27. We lack a clearly defined goal for global climate policy.

The highest factor score was +1 and the average –0.2. One could conclude that even though participants did not consider the temperature goal to be sufficiently ambitious, they perceived it as sufficiently clear to guide policy-making efforts.

28. Since the climate is going to change, we should be more concerned with adaptation.

Participants showed surprisingly little support for this statement, despite growing concerns in the media and in academia about the need to adapt, and despite the continual demands of negotiators from developing countries for developed countries to give them more help to do so. Only four participants ranked the statement +5 or +4. The highest factor score was +3 and the average factor score +0.8.

31. Geo-engineering can solve the global warming problem much more cost-effectively than mitigation.

Most study participants rejected the idea that geo-engineering might offer a more affordable solution to the climate change problem than mitigation might. The factor scores for this statement ranged from –1 to –3. This was a surprising result given that the geo-engineering literature often emphasized the appeal of this technological pathway due to its lower price tag in comparison to economically painful mitigation policies. But the argument that geo-engineering might buy more time for mitigation and adaptation generated greater agreement among most participants.

33. Economic growth and jobs must take priority over climate concerns.

There is a general belief among many UNFCCC negotiators and observers that the key concern of diplomats from developed countries is the protection of domestic economic interests. But statement 33 received negative scores from twenty-five of the twenty-nine Q study participants (86 percent), most of whom were members of groups 4-HL and 5-ML—high and medium emitters. All factor scores were negative (–2 to –3); only two study participants—NGO representatives associated with group 4-HL—assigned positive individual scores to this statement (+1 and +2). Similarly, there was fairly strong disagreement among study participants with statement 38:

38. Economic growth is the best solution to climate change.

The factor scores for this statement ranged from +1 to –5, with an average of –2.3.

45. My government is a very constructive player in international affairs.

The scores for this statement reflected the highly contentious character of UNFCCC negotiations and the rather negative self-assessment of those representing the states that are parties to them. The highest factor score was +2; four of the six factor scores were negative, ranging from –1 to –5.

52. Rich countries have caused the problem; consequently, they have the obligation to fix it.

This statement reflects the then-dominant interpretation of one of the UNFCCC's most important principles, which has been the source of major disagreements in the climate change negotiations: common but differentiated responsibilities (CBDR). Statement 52 generated mixed responses from study participants, with a wide score range (–3 to +5) and an average score of +0.24. Most of the individual scores ranged between –2 and +2. Only five participants strongly supported this statement, with scores between +3 and +5.

55. I believe that we will find a cooperative solution to climate change. Other issues have taken many years of negotiation, too.

Only one factor showed significant agreement with this optimistic statement, assigning it a factor score of +3; participants in all other factors scored it 0 or +1. Generally, participant negotiators seemed to have only modest confidence in their collective ability to find a multilateral solution in 2012. Now, after the adoption of the Paris Agreement, this statement would likely generate very different responses.

56. Climate change is a very depressing issue.

Participants in all but one factor scored this statement 0, which could be interpreted in at least two different ways. First, the lack of agreement suggests that the negative or threatening content of climate change–related information did not affect study participants in the same way it has affected ordinary citizens. Second, given the lack of disagreement with this statement, study participants might simply have been unsure how to respond to it. More generally, participants tended to rank emotional statements invoking fear, disappointment, or worry in the middle of the distribution. On the other hand, statements 57 and 58, which touch upon a sense of fatalism and being overwhelmed, triggered strong disagreement.

57. Climate change is simply too complex and overwhelming. It is impossible to fully understand, let alone solve, the problem.
58. It is already too late to do anything about climate change.

Factor scores for statements 57 and 58 ranged from –1 to –4 and from 3 to –5, respectively, with average raw scores of –1.8 and –2.9. Study participants actively rejected thoughts that were disempowering or that implied a lack of agency.

64. I have a hard time imagining the consequences of climate change for my community and my country.

Although the interview data offer evidence that study participants had difficulties imagining worst-case scenarios of the distant future (the year 2080), judging from the very narrow range of factor scores assigned to statement 64, close to 0 (–1 to 0), they generally did not believe they lacked imagination regarding how their home countries might be affected by climate change.

65. Sometimes gradual processes such as GHG emissions result in sudden, dramatic changes in the environment. The existence of such climate tipping points makes action even more urgent than previously thought. Avoiding tipping points should become a key climate policy goal.

Even though the possibility of climate tipping points had received growing attention in the scientific community, participants did not find the issue to be of high importance in the negotiations in 2012. One factor stood out (Factor B, see below), assigning a score of +5 to this statement, as compared to participants in all other factors, whose scores ranged from +1 to +2.

Six Factors

Table A.4 in appendix A.8 lists all sixty-five Q study statements with their factor scores.

Factor A—Multilateralism Champions
The statements with extreme scores listed in table 6.2 point to four themes that dominated this belief system: the importance of cooperation among all states within the UNFCCC, a focus on adaptation, economic growth as a problem rather than a solution, and the character of climate change as a moral problem.

States were the central actors among participants having the factor A belief system. One of the most important elements of global climate governance is the cooperation of all governments. Factor A participants were the only participants who strongly agreed with statement 42 ("The BASIC countries (Brazil, South Africa, India, and China) should show greater leadership in international climate change negotiations": +4), with one of them commenting, "BASIC countries need to commit more to taking action, not just leadership." Although between 2009 and 2011, the BASIC Group was an active negotiation bloc, it has since moved into the background, with very limited and general interventions during the negotiation process. China and India prefer to express their positions through a new negotiation group—the Like-Minded Developing Countries on Climate Change (LMDC).

Participants holding the factor A viewpoint strongly rejected the argument that climate change concerned only the developing countries (statement 6: –5) and that all efforts and resources to deal with climate change should be allocated domestically rather than internationally (statement 13: –4). They clearly favored a multilateral approach. Their focus on cooperation among all states indicated a strong preference for the UNFCCC

Table 6.2

Extreme factor scores for factor A

Score	Statement
+5	9. An effective climate solution requires the cooperation of all governments around the world.
	22. A key element in solving the climate problem is the need for fundamental value change within our societies.
+4	21. Neither states nor markets nor civil society can solve this problem on their own—climate change is a multilevel problem and requires action at all of these different levels.
	29. Adaptation and mitigation are complementary and equally important policies.
	42. The BASIC countries (Brazil, South Africa, India and China) should show greater leadership in international climate negotiations.
	48. All states have a moral responsibility to contribute to a global climate solution.
−5	6. Climate change is mainly an issue for the developing countries.
	38. Economic growth is the best solution to climate change.
−4	4. I do not believe that we will see significant effects of climate change in my lifetime.
	12. Only a small number of countries with significant GHG emissions are important for climate negotiations.
	13. All our efforts and resources to combat climate change should be concentrated domestically, rather than internationally.
	47. Problems that might arise decades from now are not important to me.

as a governance venue and a belief in the primacy of the multilateral level in a multilevel governance system (statement 21: +4).

Factor A participants emphasized the moral responsibility of all states to act on climate change, locating the reason for cooperation in the realm of morality and justice rather than of economic costs and benefits. Because they believed the subjects of rights and obligations were states rather than individuals, they assigned statement 49 ("The current generation (of politicians and voters) has a major ethical responsibility to future generations") a lower score (+2) than all other participants did. But holding this state-centered viewpoint, they disagreed with the notion that the causal and impact asymmetry between the rich and the poor was unjust (statement 52: −2).

Emphasizing the equal importance of adaptation and mitigation (statement 29: +4), factor A participants believed that both the causes and the consequences of climate change should be dealt with at the same time. And emphasizing adaptation, they signaled strong concern about the impacts of climate change on vulnerable developing countries, finding the social impacts, such as suffering from food and water scarcity, to be most worrisome (statement 5: +3). At the same time, they expected climate change impacts in their own lifetimes (statement 4: –4) and cared strongly about events that would occur decades into the future (statement 47: –4). Participants having this set of beliefs solidly supported "climate funding" (statement 51: +3). Combining these convictions, they had a strong sense that climate change would affect our lives in ways that required adaptation, and that it was the moral responsibility of states to prevent future climate change–related suffering with investments in both mitigation and adaptation. Their comparatively low disagreement (–1) with statement 31 on the cost-effectiveness of geo-engineering suggests that geo-engineering might be a serious policy option for this group.

Another interesting feature of factor A participants was their strong disagreement with statement 38 ("Economic growth is the best solution to climate change": –5), which suggests they saw economic growth as part of the problem rather than as the solution to climate change. Emphasizing the need for value change within our societies (statement 22: +5), participants holding this viewpoint might argue that a departure from our current growth-based development models was needed and would require a fundamental shift in beliefs about the good life, progress, modernity, and happiness, away from consumption and materialism. Along the same lines, factor A participants assigned a low score (–3) to statement 15 ("The climate problem should be left to the markets"), and the lowest of all factor scores to statement 36 ("The main costs of climate change policies include loss of GDP and jobs": –3). They considered progress on global climate governance more important than economic growth.

Finally, it is worth noting that factor A participants assigned statements 1 and 2 more moderate scores than other participants did (+3 and –2, respectively). This might indicate that the beliefs reflected in their responses to the two statements had become normal for these participants and no longer required any emphasis.

Emotions did not play a strong role in the factor A belief system. Excitement about a cleaner economic future (statement 25: +3) and disappointment with a lack of environmental concern (statement 44: +2) were the strongest expressions of emotion among factor A participants. Another interesting observation concerns the higher-than-average scores these participants assigned to statements 57 ("Climate change is simply too complex and overwhelming. It is impossible to fully understand, let alone solve, the problem": –1) and 58 ("It is already too late to do anything about climate change": –3). This suggests a greater tolerance for admitting the inadequacy and possibly the failure of climate governance efforts, although it still signaled disagreement with these statements. In line with such a willingness to accept failures and shortcomings, participants who shared the factor A belief system disagreed with statement 26 (–3) more than other participants did, indicating their dissatisfaction with the temperature goal of 2°C.

Study participants whose Q sorts loaded significantly on factor A included diplomats in groups 2-MH, 4-HL, and 5-ML and members of various nonstate actor groups, including a youth NGO and the fossil fuel industry. All but two factor A participants lived in the developed world. More generally, the themes of this belief system—centrality of state agency, the primacy of the multilateral governance level, equal weight on mitigation and adaptation, concern about the future and moral obligations of all states—resembled those identified in the CAMs of group 2-MH. Skeptical of today's model of development, factor A participants advocated changing the current way of life and consumption patterns. They also easily identified with identity groups larger than their own states—humanity or the community of states—which corresponded to the belief that all states had a moral obligation to act, and not just a select few based on their historical emissions or any other measure. Factor A participants' beliefs about state agency in a multilevel governance system also reflected parts of the cognitive patterns identified in the CAMs of group 4-HL.

Factor B—UN Skeptics
Participants having the factor B belief system attributed hardly any importance to states in global climate governance; among the statements receiving the highest and lowest scores from these participants (see table 6.3),

Table 6.3

Extreme factor scores for factor B

Score	Statement
+5	1. Human-released greenhouse gases are causing significant climate change.
	65. Sometimes gradual processes such as GHG emissions result in sudden, dramatic changes in the environment. The existence of such climate tipping points makes action even more urgent than previously thought. Avoiding tipping points should become a key climate policy goal.
+4	21. Neither states nor markets nor civil society can solve this problem on their own—climate change is a multilevel problem and requires action at all of these different levels.
	25. The prospect of a cleaner, eco-friendly economy is exciting.
	34. Ideally, climate policies would reduce GHG emissions while stimulating economic growth.
	49. The current generation (of politicians and voters) has a major ethical responsibility to future generations.
–5	2. I don't trust what scientists say about climate change.
	47. Problems that might arise decades from now are not important to me.
–4	4. I do not believe that we will see significant effects of climate change in my lifetime.
	6. Climate change is mainly an issue for the developing countries.
	24. God has made us stewards of the Earth, giving us the ability and responsibility to keep the planet healthy.
	58. It is already too late to do anything about climate change.

only one concerned state governments as agents of climate governance (statement 10: –1). Instead, the most important themes of this belief system were a strong grounding in facts and science, a concern about the future and the possibility of climate tipping points, and a struggle to balance the need to address climate change and the need for economic growth.

The extreme rankings of statements 1 (+5) and 2 (–5) demonstrate that participants holding the factor B viewpoint considered scientific knowledge foundational—factual statements were more important than value-based statements. This strong trust in data found expression in three additional ways. One was a concern about climate tipping points, which was in 2012 based purely on scientific theories and publications, not on observation or experience. The most unusual feature of factor B participants was their assigning statement 65 a score of +5; participants in all other factors ranked this statement +1 or +2. In addition to signaling trust in scientific knowledge, the score of +5 indicated factor B participants' concern about the future and their desire for long-term stability and predictability.

The second way in which factor B participants showed their preference for facts over values was their strong disagreement with statement 24 ("God has made us stewards of the Earth, giving us the ability and responsibility to keep the planet healthy": –4). Rather than ignoring this statement as not useful or not important (e.g., assigning it a score of 0), they decisively disagreed with it, suggesting their strong discomfort with giving religion or spirituality a role in the climate change debate.

Third, factor B participants ranked the need for value change in our societies (statement 22) far lower (0) than participants in any other factor did, suggesting they believed value change to be irrelevant in addressing climate change. At the same time, the slightly or moderately positive scores they assigned to statement 62 ("The focus on winning the next election is the biggest obstacle to finding international agreement": +1) and statement 18 ("Politicians need much stronger voter support and pressure from political movements to create meaningful for climate policies": +2) suggests they believed voters simply needed to understand the problem better in order to make better election choices.

Like their factor A counterparts, factor B participants strongly believed that climate change was a multilevel governance problem that could not be solved by states, markets, or civil societies alone. But they did not believe that states were the most important players in global climate change politics (statement 10: –1), or that subnational state actors had the ability to address the problem (statement 16: –1). Instead of cooperation among all states (statement 9: 0), factor B participants believed that only a small group of states was sufficient to address climate change at the multilateral level (statement 12: +2), presumably to provide a global framework for mitigation. Rather than pursuing negotiations in the UNFCCC, they preferred a "club" approach. They also disagreed less than participants in other factors did with the argument that efforts to fight climate change should be concentrated at the domestic rather than international level (statement: 13: –1). Depending on a factor B participant's nationality, this could be an argument for more ambitious domestic mitigation policy or for the need to push adaptation at home. On the other hand, it could be interpreted as a lack of support for climate change finance, technology transfer, and other international support measures for the vulnerable and poor. Finally, factor B participants clearly disagreed with the argument that elected officials had to protect the interests of the present rather than future generations (statement 46: –3), and they even more clearly agreed that elected officials had a major ethical responsibility to future generations (statement 49: +4). To summarize, those having the "UN skeptics" belief system advocated a club approach at the international level and placed the key burden of responsibility for climate governance on elected officials in a national political context.

For factor B participants, the key challenge of global climate governance was the reconciliation of mitigation and growth—ideally, climate change policies would reduce GHG emissions while stimulating economic growth (statement 34: +4). But they neither agreed nor disagreed with the argument that economic growth and jobs had to take priority over climate change concerns (statement 33: –3). Assigning statement 33 a score of –3, they did not consider economic growth to be a solution to the climate change problem or an end in itself, but more a necessity. In contrast to factor A participants, individuals holding the factor B point view did not call for a new development model, but did believe in the need for development. They were excited about the prospect of a cleaner economy (statement 25: +4), but they did not trust the problem-solving power of the markets (statement 15: –2). Their policy goal was a balance between the climate change policies and the economy, and they were likely to support green growth and sustainability concepts. Although they did not believe states were the most important actors in climate change politics, they did seem to believe states had a role in creating the regulatory frameworks that pursued achieving a balance between emission reductions and economic growth.

Participants holding the factor B viewpoint cared strongly about problems that would occur decades into the future, and they expected to experience impacts of climate change in the course of their lifetimes. They strongly disagreed that it might already be too late to do something about the problem (statement 58: –4), suggesting they believed that actors in 2012 were able to prevent future harm, including the occurrence of climate tipping points. At the same, they were fairly pessimistic about the prospects for major change in the near future, assigning statement 20 ("Nothing will happen before a climate crisis hits") a higher score (+1) than participants in any other factor did.

With the exception of statement 25 ("The prospect of a cleaner, eco-friendly economy is exciting": +4), emotional statements were not very important for factor B participants. They had more fear for their children's future (statement 54: +2) than participants in all other factors except D did, but no uncertainty, no sense of being overwhelmed (statement 57: –3), and little concern about having to explain the failure to address climate change to their grandchildren (statement 63: –1). In other words, these individuals experienced no personal shame or guilt for not having lived up to their responsibility.

Study participants with significant loadings on factor B included members of groups 2-MH, 4-HL, 5-ML, 6-LL and a youth NGO representative. Members of groups 4-HL and 5-ML were in the majority. Factor B participants shared a number of cognitive patterns with those identified among industry NGOs associated with group 4-HL: an extraordinary concern for climate tipping points and an ability to deal with the fairly long time horizons of climate change, a belief in the important role for governments domestically rather than internationally, a preference for market-based approaches but not laissez-faire policies (e.g., establishing a carbon price signal through carbon markets), and a belief in the need for greater voter support for politicians in favor of action on climate change. The view that the UNFCCC negotiations were less important than the numerous climate change efforts outside of them, ranging from small club and bilateral diplomacy to subnational and NGO initiatives, was prominent among the diplomats in group 4-HL. Members of group 5-ML—both diplomats and environment and market NGO representatives— also expressed a notable concern about climate tipping points.

Factor C—Utilizing the Market

Although there are some significant similarities between the extreme scores of factor B and C participants (see tables 6.3 and 6.4), when combined with the scores factor C participants assigned to other statements, these extreme rankings provided the foundations for a distinct belief system that emphasized state cooperation within the UNFCCC, links between international and domestic politics, a strong concern with economic growth, and a personal element that links their identity with climate change.

Table 6.4

Extreme factor scores for factor C

Score	Statement
+5	1. Human-released greenhouse gases are causing significant climate change.
	48. All states have a moral responsibility to contribute to a global climate solution.
+4	9. An effective climate solution requires the cooperation of all governments around the world.
	18. Politicians need much stronger voter support and pressure from political movements to create meaningful climate policies.
	21. Neither states nor markets nor civil society can solve this problem on their own—climate change is a multilevel problem and requires action at all of these different levels.
	25. The prospect of a cleaner, eco-friendly economy is exciting.
–5	2. I don't trust what scientists say about climate change.
	39. Investment in climate policies is a poor use of our resources; it makes more sense to do something about poverty, health care and education in the developing world.
–4	6. Climate change is mainly an issue for the developing countries.
	13. All our efforts and resources to combat climate change should be concentrated domestically, rather than internationally.
	47. Problems that might arise decades from now are not important to me.
	58. It is already too late to do anything about climate change.

Factor C participants' high scores for statements 1 (+5) and 2 (–5) grounded their belief system in facts and science, similar to the belief system of factor B participants. But factor C participants assigned the highest score also to statement 48 ("All states have a moral responsibility to contribute to a global climate solution": +5).

Factor C participants also emphasized the multilevel nature of the climate change problem (statement 21: +4) and the essential role of multilateral negotiations in its solution (statement 9: +4). But because they believed that the concern of some UNFCCC delegations and political actors about the next domestic election was an obstacle to finding an international agreement, they also believed that politicians needed "much stronger voter support and pressure from political movements" (statement 18: +4). Factor C participants assigned statement 19 on the blocking power of vested interests the lowest score among participants in all factors (–2). Their disagreement with this statement could mean either that they considered vested interests not to have much blocking power, or that this power could be overcome with voter support and grassroots pressure.

A third theme among factor C participants was their strong preference for investments in climate change policies rather than in alleviating poverty or promoting health care and education (statement 39: –5), and for doing so at the international rather than domestic level (statement 13: –4). Combined with their rejection of statement 6 ("Climate change is mainly an issue for the developing countries": –4), this suggests they strongly believed in global cooperation and that the climate change problem could not be addressed without some form of global governance. This was not, however, a call for international support schemes for poor countries, but a recognition that effective mitigation required global cooperation.

Participants having the factor C belief system showed significantly less concern than other participants about the social consequences of climate change (statement 5: 0). They assigned the possibility of death and violence as a consequence of climate change (statement 8) the lowest factor score (+1), as they did unfairness concerns (statement 7: +1). Their slight disagreement with statement 28 (–1), suggests a stronger focus on adaptation, and they were the only participants who did not fully support "climate funding" (statement 51: 0).

Participants who shared this belief system believed in the market as a source for solutions more than other participants did (statement 15: +1). They were excited about a cleaner economy (statement 25: +4), and they were the only participants to agree with statement 38 ("Economic growth is the best solution to climate change": +1). Their other scores confirmed their use of economic frameworks of analysis when thinking about climate change: they supported cost-benefit analysis (statement 35: +2), and they scored statement 36 ("The main costs of climate change policies include loss of GDP and jobs") higher (0) than participants in the other factors did, even as they also agreed with statement 37 ("The main costs of future climate change can simply not be calculated: the loss of human life, food insecurity, or species extinctions don't have price tags": +2). The relationship between these scores suggests that the factor C participants' belief system was firmly rooted in economic frameworks, but that other values occasionally challenged or even overrode this economic orientation.

Participants loading on factor C seemed to view climate change as an issue in which they were personally—not only professionally—invested. They dedicated their careers to solving the climate change problem (statement 43: +3), and they expressed greater optimism than other participants did in the possibility of finding a solution through negotiations (statement 55: +3). They had a strong sense of agency, rejecting the feeling of helplessness (statement 53: –3). Factor C participants were also one of only two factor groups to signal some concern about having "to explain our failure to fix the climate problem" to their grandchildren (statement 63: +1), connecting agency and future accountability for failure.

Their personal perspective seemed to be linked to a strong sense of national identity. Factor C participants perceived states to be "important players in global climate politics" (statement 10: +1), and they strongly disagreed with statement 41 ("I am ashamed that my country is not doing more about climate change": –3). They disagreed more than other participants did with the idea that the rich countries had caused climate change and therefore had an obligation to fix it (statement 52: –1), and strongly agreed with statement 48 ("All states have a moral responsibility to contribute to a global climate solution": +5), although they may have ranked this statement so highly because of its emphasis on all states, rather than on moral obligations, which might reflect their desire to move themselves out of the spotlight of responsibility.

Factor C participants had unusually strong negative responses to statements that referred to emotions. Apart from their rejection of shame (statement 41: –3), they were not particularly upset about the power disparities between the rich and poor (statement 14: –2), nor particularly disappointed about the lack of environmental concern, assigning statement 44 the lowest score of any participants (0). Their rejection of helplessness (statement 53: –3) was coupled with an ambivalent score about the loss of hope (statement 59: 0).

Members of groups 4-HL and 5-ML and one individual in group 2-MH shared the factor C viewpoint. All but one were diplomats. The patterns identified here match those identified with the CAM analysis of members of group 4-HL: a focus on the costs of action rather than the social costs of climate change impacts, a concern with domestic political constraints affecting multilateral negotiations, a preference for global cooperation combined with confidence in market forces to contribute to a solution (e.g., the UNFCCC's Clean Development Mechanism), and the absence or lack of importance of norms of justice.

Factor D—The Power of Individuals

Instead of being concerned with states or market forces, participants having the factor D belief system placed heavy emphasis on individuals and their political roles and power to create change (see table 6.5 for extreme factor scores). Skeptical of the ability of economic growth, market forces, and neoliberal policy frameworks to address the climate change problem, factor D participants acknowledged their responsibility to future generations, decisively disagreed that geo-engineering would be more cost effective than mitigation in solving the climate change problem, and they believed the developed world had an obligation to act on climate change.

Table 6.5

Extreme factor scores for factor D

Score	Statement
+5	18. Politicians need much stronger voter support and pressure from political movements to create meaningful climate policies.
	22. A key element in solving the climate problem is the need for fundamental value change within our societies.
+4	1. Human-released greenhouse gases are causing significant climate change.
	7. Climate change was caused by rich, industrialized countries, but its impacts will hurt poor countries most—this asymmetry is unfair.
	21. Neither states nor markets nor civil society can solve this problem on their own—climate change is a multilevel problem and requires action at all of these different levels.
	49. The current generation (of politicians and voters) has a major ethical responsibility to future generations.
−5	6. Climate change is mainly an issue for the developing countries.
	15. The climate problem should be left to the markets.
−4	2. I don't trust what scientists say about climate change.
	47. Problems that might arise decades from now are not important to me.
	50. Future generations are likely to be richer and better off than we are, and better able to deal with climate change.
	58. It is already too late to do anything about climate change.

Believing as well that the key for addressing climate change lay in the values and attitudes of individuals, who had the power—as voters or grassroots organizers—to change the politics of climate change and to influence their government's negotiation position, they, alone, assigned the highest factor score (+5) to statement 18 and, with factor A and F participants, to statement 22, rejecting just as strongly statement 15 ("The climate problem should be left to the markets": –5). In line with their focus on individual agency, factor D participants disagreed with statement 17, according to which individual contributions do not make a difference (–3). Value change at the individual level is the first step toward meaningful action on climate change at the community, national, and international levels, but it is still unclear what values should be discarded and which new ones adopted. The unusually high score (+3) factor D participants assigned to statement 23 ("Based on our shared humanity, our desire for happiness and security, we can find a solution to the climate problem") suggests they might find collective human values like happiness and security rather than individual ones like independence and property to be the values to adopt.

Clearly perceiving the asymmetry of climate change impacts between rich and poor to be unfair (statement 7: +4), factor D participants were upset about power disparities among countries in the UNFCCC negotiations (statement 14: +2), strongly rejecting the idea that climate change is a problem for the developing countries alone (statement 6: –5), which suggests they believed that leaving the developing countries—in essence, the victims—to deal with this problem on their own would be immoral and that developed countries had a moral duty to address climate change. But the high score (+4) they assigned to statements 7 and 49 also suggests they believed this duty was humanity's, rather than the states'. They neither agreed nor disagreed with the argument that states were the most important players in global climate change politics (statement 10: 0); they assigned the lowest factor score of all participants (+2) to statement 48 on the moral duty of all states to contribute to solving the climate change problem; and they only slightly agreed with statement 52 ("Rich countries have caused the problem; consequently they have the obligation to fix it": +1). All of which suggests factor D participants believed that, though states mattered somehow, it was individual human beings who had the moral duty to address climate change.

The ethical framework of the factor D belief system had a number of extensions. The most obvious one was its participants' concern about future generations, who were expected to be worse off than the 2012 generation due to climate change. Their belief that their generation played a role in diminishing the potential for happiness of future generations added to a sense of personal responsibility among factor D participants. This was expressed in at least three ways: their disagreement with the idea that problems that might arise decades into the future were not important (statement 47: –4), fear for their children's future (statement 54: +2), and concern about having to explain their failure to address climate change to their grandchildren (statement 63: +1), with scores on these last two statements that were higher than those of all but one other factor group.

Another extension was a fairly strong rejection of neoliberal ideas that prioritize economic growth (statements 33 and 38: both –3), that utilize market forces to address the problem (statement 15: –5), and that frame the climate change issue in terms of a cost-benefit analysis (statement 35: –2). Instead, factor D participants believed that vested interests were "too powerful to allow for any meaningful action on climate change" (statement 19: +2).

This ethical framework and distancing from neoliberal ideas may also have been linked to factor D participants' rejection of geo-engineering, either as a cost-effective solution to the mitigation challenge (statement 31: –3) or even as a tool to buy more time for mitigation and adaptation (statement 30: –2). They might have had an underlying mistrust of technological quick fixes that merely sought to avoid dealing with their generation's ethical responsibilities. Geo-engineered solutions might have allowed the world to continue on its current economic development path without the value change that factor D participants identified as most important for addressing climate change (statement 22: +5).

Factor D participants assigned fairly high factor scores to five emotional statements: excitement about a cleaner economic future (statement 25: +3), disappointment about a lack of environmental concern in state and market institutions (statement 44: +3), distress over major environmental change (statement 40: +2), fear for their children's future (statement 54: +2), and upset about power imbalances in the UNFCCC (statement 14: +2). With the exception of fear for their children's future, all of these were emotional responses to collective states of affairs or societal conditions.

The ten Q study participants whose sorts loaded significantly on factor D represented all six participant groups. They included diplomats from groups 2-MH, 3-LH, 4-HL, and 6-LL and NGO representatives from groups 1-HH, 2-MH, and 5-ML. Given this diversity of individuals sharing the same viewpoint, it is difficult to identify a participant group whose CAM patterns offer a strong match with the cognitive patterns of factor D, which presents a combination of beliefs that exist across different participant groups.

The belief that individuals matter as voters in a two-level game showed up consistently across all six participant groups, among diplomats and NGO members alike. Participants who believed in the power of voters also tended to emphasize the power of civil society organizations to change voter attitudes, to create the broad public awareness through framing and mobilization that allowed governments to take costly action on climate change and adopt more ambitious negotiation positions in the UNFCCC. This focus on individuals and civil societies was particularly strong among members of group 2-MH, representatives of environment and market NGOs (group 5-ML), and diplomats in group 6-LL. The diplomats in group 2-MH identified additional ways in which individuals could make a difference: decisions by political leaders and lifestyle changes by consumers. Youth NGO members added another aspect to individual responsibility, emphasizing the importance of local and community activities through which people could take care of one another and take responsibility for the environment.

Certain beliefs among Youth NGO members supported two additional features of the factor D belief system: they had a desire to speak on behalf of future generations and were concerned about the implications of climate change for those future generations. They also identified with young people around the world, especially with members of the climate change youth movement, emphasizing shared elements of human identity that went far beyond the state.

The idea of shared human values and a human community had a strong influence on the beliefs of groups 2-MH, 3-LH, and 6-LL. They emphasized values like mutual survival, solidarity, and protection of the weakest, and argued that ethics should take priority over profits. In their view, inaction regarding climate change was a moral failure inconsistent with the values of the human community.

Factor E—Climate Justice

The extreme scores of participants having the factor E belief system (see table 6.6) showed major overlap with those of the factor D participants. The similarities included a heavy emphasis on unfairness, which was even more pronounced in factor E participants, a belief in the power of voters and political movements to influence global climate change politics indirectly by creating political pressure on politicians at the domestic level, the ethical responsibility of the current generation of politicians, who must not leave the developing countries to grapple with climate change on their own, and a lack of confidence in the power of the market to solve the climate change problem. However, there were also a number of significant differences between these two belief systems.

First, the argument that rich countries had an obligation to fix the problem because they had caused it in combination with an emphasis of

Table 6.6

Extreme factor scores for factor E

Score	Statement
+5	1. Human-released greenhouse gases are causing significant climate change.
	7. Climate change was caused by rich, industrialized countries, but its impacts will hurt poor countries most—this asymmetry is unfair.
+4	18. Politicians need much stronger voter support and pressure from political movements to create meaningful climate policies.
	21. Neither states nor markets nor civil society can solve this problem on their own—climate change is a multilevel problem and requires action at all of these different levels.
	49. The current generation (of politicians and voters) has a major ethical responsibility to future generations.
	52. Rich countries have caused the problem; consequently they have the obligation to fix it.
−5	6. Climate change is mainly an issue for the developing countries.
	58. It is already too late to do anything about climate change.
−4	2. I don't trust what scientists say about climate change.
	15. The climate problem should be left to the markets.
	7. Problems that might arise decades from now are not important to me.
	57. Climate change is simply too complex and overwhelming. It is impossible to fully understand, let alone solve, the problem.

unfairness suggest that this perspective was about climate justice as many developing country diplomats and members of the climate justice movement then defined it. Confirming this assumption, the participant with the highest loading on factor E stated that she had assigned score +5 to statements 7 and 52 because "climate change is a climate justice issue" (her factor group gave the two statements scores of +5 and +4, respectively). This perspective also explains factor E participants' strong disagreement with statement 6 (–5). She commented, "Developed countries have the biggest share of responsibility in dealing with the problem." This view was strongly grounded in a distinction between developed and developing countries and placed responsibility for action on climate change with the developed world. States played an important role, either as perpetrators or victims of climate injustice.

Second, the high score factor E participants assigned to the ethical argument in statement 49 (+4) was likely intended to emphasize the responsibility of politicians rather than individual voters to take action on climate change, a likelihood bolstered by the comparatively low score they assigned to statement 22 ("A key element in solving the climate change problem is the need for fundamental value change within our societies": +1). Whereas factor D participants gave statement 22 their highest ranking (+5), factor E participants de-emphasized the role of individuals and civil societies while elevating that of state governments.

Third, factor E participants strongly disagreed with the suggestion that the climate change problem might be too complex to understand or solve (statement 57: –4), which suggests they believed that somebody or some group of actors was capable of solving the climate change problem with the knowledge and technologies existing in 2012. The strength of their rejection of statement 57 suggests a concern that the rich could use this argument as an excuse for not acting and for shirking their moral obligations.

Fourth, factor E participants emphasized the importance of adaptation and support for the developing countries (statement 28: +3), a view not shared by their factor D counterparts (–1). Factor E participants expressed significant support for the argument that economic growth was desirable (statement 34: +3). Given their views on the differences between the global North and South, they might have intended their support for growth to apply to the developing world only. At the same time, they rejected the idea that economic growth was a "solution to climate change" (statement 38: –3). Further, they strongly disagreed with the suggestions that taxes or other property constraints were politically unacceptable (statement 32: –3). As one factor E participant commented:

The enforcement of property rights is one of the main obstacles to ... progress on the climate issue—they prevent the transfer of what would be extremely useful and effective technologies, finance, and intellect.... Since the purpose of politics is to serve society, it should be irrelevant whether this responsibility is politically popular or not.

Participants who loaded on factor E included members of youth NGOs, and groups 2-MH, 4-HL, and 5-ML. The strongest match between CAM patterns and the beliefs of factor E existed for group 2-MH, especially faith and development NGOs, who made a clear distinction between developed and developing countries, used a justice framework to explain the relationship between them, saw historical responsibility as the foundation for the obligation of the developed world to address the climate change problem, supported adaptation measures and resource flows from North to South, and focused on state actors rather than civil societies, markets, or individuals as principal agents for action on climate change.

Factor F—Spotlight on the West

Five themes characterized this final belief system (see table 6.7). A first and the most important theme was the need for value change within our—presumably Western—societies as an ethical responsibility of the current generation to future ones. The future was an important part of factor F participants' perspective, a second theme: they expected it to be worse than the present due to climate change impacts in their own lifetimes. A third theme concerned the importance factor F participants attached to national identity and their associated feelings of disappointment and shame about their governments' poor records on climate change. Weaker fourth and fifth themes were their being upset about existing power disparities in the negotiations, which hinted at their underlying equity concerns and focus on the North–South divide, and their skepticism about the market's power to address climate change.

Value change was at the center of the factor F belief system. When acted on in elections or in civil society organizations, value change can influence national politics and eventually international negotiations. Factor

Table 6.7

Extreme factor scores for factor F

Score	Statement
+5	22. A key element in solving the climate problem is the need for fundamental value change within our societies.
	49. The current generation (of politicians and voters) has a major ethical responsibility to future generations.
+4	14. The power disparities between countries in climate negotiations make me very upset.
	18. Politicians need much stronger voter support and pressure from political movements to create meaningful climate policies.
	25. The prospect of a cleaner, eco-friendly economy is exciting.
	41. I am ashamed that my country is not doing more about climate change.
−5	45. My government is a very constructive player in international affairs.
	47. Problems that might arise decades from now are not important to me.
−4	2. I don't trust what scientists say about climate change.
	4. I do not believe that we will see significant effects of climate change in my lifetime.
	15. The climate problem should be left to the markets.
	50. Future generations are likely to be richer and better off than we are, and better able to deal with climate change.

F participants' comments and CAMs elucidated why they believed value change in our societies (statement 22: +5) was so central to addressing climate change and what type of value change they envisioned. As one factor F participant explained:

I think a new political wave ... worldwide presents a potential to solve the climate problem, and I think that an enhanced understanding of human connectedness as well as ecosystem and earth system sciences play an important role in creating a foundation for principled climate politics. When I think about the need for value change, or "a new story" to define climate politics, I am also thinking about the need to expand collective consciousness (or identify) beyond individual, family, and national borders.

The same individual justified assigning the second highest factor score to statement 18 ("Politicians need much stronger voter support and pressure from political movements to create meaningful climate policies": +4):

I do not see it possible to realize significant progress on climate change without a strong mobilization for change and action from local, provincial, and national governments. It is necessary for many complementary efforts to unfold at the same time, ranging from market measures, innovation, and enhanced communication of science. But if state policies are chosen with the goal of ensuring an elected party's political survival, then it is incumbent on the electorate to demand strong climate policy.

The CAMs of individuals with high loadings on factor F contained concerns about the growing consumption orientation of the world, the need to rethink the old development paradigm of the West, including "what we eat, what we produce, and where our ambitions lie," the need for more community focus, sustainable lifestyles, and emotional health rather than "North American consumerism and entertainment focus," and a "social conscience shift" in the next generation of voters, which required "special people" to take risks and assume leadership. Addressing climate change required personal change by individuals in Western societies, who, on the one hand, needed to reorient their personal and community lives and, on the other, had to exercise their democratic rights and push their societies in a new direction with their votes and political mobilization for action on climate change. But factor F participants neither agreed nor disagreed that "our shared humanity" (statement 23: 0) would play a role in this reorientation.

These individuals only slightly disagreed with statement 17 ("Individual contributions don't make a difference when it comes to climate change.": –1), suggesting they believed these contributions had only a limited effect if they were not complemented by actions at other levels of civil society. The factor scores assigned by factor F participants to statements 7 (+3) and 15 (–4) suggest they believed that "rich industrialized countries" rather than "the markets"—had a strong role to play in solving the climate change problem. Opposing the idea that climate change was "mainly an issue for the developing countries" (statement 6: –3), believing that all states have a moral responsibility to contribute to a solution (statement 48: +3), and giving the highest of all factor scores to the argument that states are the most important players in global climate change politics (statement 10: +2), participants holding the factor F viewpoint focused on the big emitters in the Global North and their responsibility to deal with mitigation. Apart from unfairness concerns and the need to address them (statement 7: +3; statement 14: +4; statement 51: +3), they did not assign high factor scores to issues related to the North–South divide. Factor F participants were ambivalent about the UNFCCC as a forum to create the necessary changes, assigning the highest factor score for statement 59 ("There are moments when I lose all hope that the UNFCCC process can solve this problem": +2).

Like their factor E counterparts, factor F participants emphasized the importance of the future and were pessimistic about the trajectory of humanity. Like them, they expected that future generations would be worse off and that climate change impacts would occur in their lifetimes. Individuals who shared the factor F viewpoint joined their factor B counterparts (see table 6.3) in assigning the highest score among all participants to the idea that climate change would "result in violence and human deaths" (statement 8: +3), and they also agreed that the "social consequences of climate change" were "most worrying" (statement 5: +2), but they disagreed that climate change was just one among many important problems today (statement 61: –2).

Although not trusting the markets to deal with climate change, factor F participants were not opposed to the use of market-based mechanisms or to the mobilization of private resources for action on climate change (statement 21: +1). As another factor F participant commented:

I think climate markets play an important role in solving the climate problem.... I also believe in the potential for market-based measures to generate revenue to support public goods, such as mitigation and adaptation measures in developing countries, while also removing incentives to pollute. However, I think the scope of the climate problem is so huge that its solution will not be ushered in by market logic, especially when the growth imperative that drives state interests is also closely coupled with carbon emissions.

Factor F participants ranked emotional statements higher than participants in any other factor did (statement 14: "upset"; statement 25: "exciting"; and statement 41: "ashamed" were all ranked +4). Particularly noteworthy is the high score they assigned to "ashamed," which had a large score spread (–3 to +4) and an average raw score of 0. Factor F participants were the only ones able to embrace being ashamed as a negative self-referential emotion, facilitated by national identity as the link between themselves and the shameful behavior of their states. These individuals felt ashamed on behalf of their governments, which had disappointed their expectations or norms of behavior. They also assigned a fairly high score (+2) to disappointment (statement 44), hopelessness (statement 59), and depression (statement 56); their perception of climate change as depressing was an exception among Q study participants in other factors, all of whom gave statement 56 a score of 0.

The factor F belief system was shared exclusively by NGO representatives, including individuals in the youth, faith, environment, and business communities. No single participant group offers a good match in terms of the cognitive patterns identified in the CAMs and the factor interpretation. This is a belief system that my CAM analysis was not able to identify in its entirety.

Comparative Assessment

Each of the six belief systems had a unique cognitive structure, focusing on different actor types, the sources of these actor types' respective obligations to address climate change, the shape their action should take, and the structural constraints they had to contend with, the prospects for success and the emotions associated with these beliefs.

At the center of each belief system was an actor type that grounded, facilitated, and constrained all other beliefs, for example, states or individuals. The centrality of actor concepts was unsurprising, yet worth emphasizing. Without clearly defined actors, it is impossible to think about a problem in terms of threats or opportunities, about obligations or possible actions. Everything hinges on the existence of actors, their characteristics and capabilities. Study participants for most of the six belief systems acknowledged different actor types, but focused on one in particular to create a coherent set of beliefs. That actor became the center of the participants' belief system; everything else made sense relative to it. Participants' meaning making required, first of all, identifying an actor and possibly their own identification with that actor or actor group.

For example, for factor A participants ("Multilateralism champions"), a focus on the international community as a collective actor, was associated with a cluster of concepts that defined the identity of this group. This cluster included concepts associated with the growth-oriented model of development that was then shared by most states, and with an ethical framework that placed states at the center. Factor A participants developed solutions for climate change from the perspective of the international community of states, bearing in mind the specific capabilities and constraints of this particular actor group, for example, its ability to create multilateral agreements and its struggle to overcome North–South differences at the time.

At the other end of the actor spectrum was the individual, the defining feature of the factor E belief system ("The power of the individual"). With identity concepts and ethical norms focused on the individual, factor E participants were involved in creating a cosmopolitan community of individuals around the world, rather than in thinking about a community of states. From this starting point, their options for action on and solutions to climate change depended on the ability of individuals (alone or collectively) to influence the political process as voters, consumers, or community organizers.

A state-based belief system, factor C ("Utilizing the market") stood out because it did not include ethical obligations for anybody to address climate change. Factor C participants had very few concepts regarding the actor's identity and character—the state as an idea having become so normalized for them that it did not even require any description. These participants strongly reacted to emotional statements, denying that emotions should play a role in thinking about climate change cooperation. Their nonmoral and antiemotional viewpoint had very little in common with the viewpoints of participants having any of the other five belief systems identified here. At the opposite end of the identity and emotional spectrum, factor F participants ("Spotlight on the West") responded to the climate change issue by embracing negative emotions, even those relating to their own national identity.

Study participants in some factors had a harder time defining the relevant actor groups than their counterparts in other factors did. For example, factor C ("Utilizing the market") participants used vague notions of market power and market mechanisms without pointing to specific actors or actions that would address the climate change challenge, although the picture became clearer once the role of the state as a regulator and creator of economic frameworks was added. Similarly, factor E ("Climate justice") participants referred to "the developed countries" or the North–South divide, and factor F ("Spotlight on the West") participants to "the Western countries" to present their viewpoints. Understanding these perspectives and their actor definitions requires considerable systemic knowledge and historical-ideational context, such as the global history of development that gave rise to concepts like the "Global South" or "Western countries." This might be a drawback for those seeking support for factor C, E, and F belief systems among populations who lack such contextual knowledge. Logical coherence is harder to achieve for people who acquire one of these belief systems.

Table 6.8 summarizes the central cognitive themes of all six belief systems—factors A through F—with an emphasis on beliefs about agency, structure, identity, justice, and the special characteristics of climate change.

Comparing these belief systems with a particular focus on their emotional content and structure is particularly interesting for a number of reasons. Most important, belief systems that are alike in terms of their emotional structure might offer more opportunities for integration and bridge building than belief systems with opposite or very different emotional profiles. Table 6.9 offers a quick glance at the profiles of the six factors.

There is broad agreement among participants in all six factors regarding a large number of emotions, although they assigned most of the negative emotions surprisingly low scores (e.g., fear driven by uncertainty about future impacts: 0 to +2). Not surprisingly, they gave the most positive score to only one positive emotion—excitement (+2 to +4). Of particular note is the near-consensus on statement 56 ("Climate change is a depressing issue"), which participants in all factors except F scored 0 (neither agree nor disagree).

Being overwhelmed by the complexity of climate change (statement 57) was the only feeling that received negative scores from participants in all factors, although the strength of rejection differed (–4 to –1). Feeling in control, in other words, not being overwhelmed by the scale or complexity of climate change and the daunting challenge of designing a functioning governance regime, was an important emotional need for participants across all factors: a sense of control was linked to agency and was an important ingredient for defining obligations and tasks of various actors. Although most people maintain a sense of control simply by rejecting a number of concepts and emotions, it is questionable whether this form of control is a productive cognitive mechanism. To the extent that control over climate change or the future of humanity is a fiction, maintaining a false sense of control or controllability could severely limit the governance options actors considered and the actions they eventually took. But participants in none

Table 6.8
Cognitive themes in Q factors

	Factor					
	A Multilateralism champions	B UN skeptics	C Utilizing the market	D The power of individuals	E Climate justice	F Spotlight on the West
Agency	States in the UNFCCC (especially BASIC)	Small group of states, elected officials	UNFCCC, states, voters	Individuals	States	Western states, voter-consumers
Structure		Two-level game, domestic politics/elections	Two-level game, domestic politics/elections	Domestic politics, elections, vested interests	Two-level game	
Identity	Community of states	National	Individual, national	Humanity	Developed vs. developing countries	National
Justice	Climate change is a moral issue, states are the subject of norms			Fairness and ethics, individuals are the subject of norms, rich vs. poor	Fairness and ethics, North–South equity and wealth distribution	North–South equity
Ideas and values	New development paradigm, growth and consumption are the problem, value change	Facts and science	Growth and market forces	Value change, shared humanity, community focus	Poverty eradication	Value change, sustainable lifestyles, voting
Emotions	Excitement, disappointment	Excitement, distress	Rejection of shame, fear for kids' future	Excitement, disappointment	Rejection of overwhelming complexity	Upset, shame, depression
Special characteristics		Climate tipping points			Rejection of overwhelming complexity	
Other	Equal focus on adaptation and mitigation; geo-engineering remains an option	Balancing mitigation and growth	Global cooperation on mitigation	Rejection of geo-engineering and neoliberalism	Emphasis on adaptation, mistrust of market mechanisms	Focus on future generations

Table 6.9

Emotional patterns in Q factors

Feeling	Factor score										
	−5	−4	−3	−2	−1	0	+1	+2	+3	+4	+5
(14) Upset				C		B	A	DE		F	
(25) Excited								E	AD	BCF	
(40) Distressed							AF	CD	BE		
(41) Ashamed				A			DE	B		F	
(44) Disappointed							BE	AF	D		
(53) Helpless			C	B	ADE	C	F	BD			
(54) Fearful (for kids' future)						AEF	C				
(56) Depressed						ABCDE		F			
(57) Overwhelmed		E	BCF		AD						
(59) Hopeless						CDE	AB	F			
(60) Scared (of unknown)					ABCE	DF					
(63) Guilty					BF	AE	CD				

of the six factors used their sense of control as a motivator for action. It is unclear how individuals engaged in the global political process should deal with this problem.

There are significant differences between participants in the six factors regarding their feelings of being upset, ashamed, and helpless. For each of these emotions, the scores of factor C and F participants were at opposite ends of the spectrum (–2 to –3 and +1 to +4, respectively), indicating that participants in these two factors had symmetrically opposed emotional patterns and were unlikely to offer each other opportunities for agreement. Participants in the other factors did not have such emotional obstacles to reconciliation. Factor F ("Spotlight on the West") was an emotional outlier that did not find acceptance among participant diplomats, possibly because its participants embraced negative self-referential emotions like shame.

Participants in factors A and D had very similar emotional patterns. Those in factors D and E and in factors A and B showed somewhat weaker similarities.

Summary

Although working with the same participants and the same topic, the Q study offered quite a different perspective on cognitive patterns among those engaged in the global climate change negotiations, compared to the CAM analysis. The Q method imposed a number of constraints that enabled greater comparability across different participants' viewpoints and facilitated a more rigorous comparison. All Q study participants had to respond to the same set of statements and thus to work with prescribed and fixed conceptual resources, which clearly limited how well they could express their idiosyncratic beliefs. For example, based on the set of statements used here, the Q analysis was not able to identify the rich diversity of beliefs regarding the different cost categories linked to climate change, which was so important for the CAM analysis, or the different identity groups participants associated with. But this exploratory weakness was balanced by the ability of the Q method to identify shared viewpoints among participants regardless of their membership in a specific group, whether participant, negotiation, or any other. The CAM analysis had to rely on artificially created categories like participant groups and was not able to uncover "natural" groups of individuals who thought and felt alike. Given their different strengths and weaknesses, both methods complemented each other well.

The Q study identified six distinct belief systems that captured a large percentage of but far from all the variance in the beliefs of the twenty-nine participants. The data contained an exceptionally high number of confounded Q sorts, which suggests that most participants had more than one belief system or combined parts of different belief systems in unique ways. This indicator of the great complexity of climate change as a global governance issue, however, also suggests bridges between different viewpoints.

The six belief systems (factors A through F) presented above reflect mainly the views of individuals in participant groups 2-MH, 4-HL, 5-ML, and youth NGO representatives. Important perspectives are not included (e.g., those of AOSIS, ALBA, LDCs) because of the constraints of the P set. The Q study was thus not able to assess how the six belief systems might be distributed among the larger population of those engaged in global climate change negotiations.

Some of the factors described above strongly matched the cognitive patterns of certain participant groups, which were the subject of the CAM analysis in chapter 5. For example, the cognitive patterns of participants in factor A ("Multilateralism champions") were very similar to those of group 2-MH; other alignments existed between the cognitive patterns of participants in factor B ("UN skeptics") and those of NGO participant representatives associated with group 4-HL, or the patterns of participants in factor C ("Utilizing the market") and those of participant diplomats in group 4-HL. For participants in the remaining factors, such a clear match did not exist. A major advantage of using the Q method in my study was the emergence of natural groups with a shared viewpoint that contained individuals from multiple participant groups. These natural groups were not groups in the sense that their members had a shared identity and coordinated their actions; indeed, it was most likely they did not know or interact with one another. Yet, unbeknownst to one another, they shared a set of beliefs and emotions concerning multilateral cooperation and climate change. The most impressive example of such a case of cognitive-affective affinity without a social group were the participants in factor D ("The power of individuals"), who belonged to all six participant groups and whose beliefs were therefore independent of national GHG emission levels or vulnerability.

Regardless of their respective factors, all study participants shared a number of beliefs concerning the reality of climate change and the trustworthiness of scientists, a concern about the future, an expectation that climate change impacts would occur in their lifetimes, and a strong conviction that it was not yet too late to address the problem. They also shared a belief in the multilevel nature of global climate governance, and the need for greater voter support and grassroots pressure, possibly combined with societal value change, to create the conditions for political action on climate change.

In 2012, study participants were strongly divided over a small number of questions: whether all states or just a small number of big emitters were necessary for an effective international agreement pitted those in factors A and C against those in factor B. The need for broad value change within our societies was contested between those in factors A and D, on the one hand, and those in factors B and C, on the other. Whether market forces could address climate change placed participants in factors C and F at odds with participants in all other factors. Contested among participants in all six factors was whether a moral obligations framework was more appropriate for dealing with climate change than a national interests framework. Participants in factors A, E, and F favored the former; those in factors B, C, and D the latter. This reflects the CAM-based finding that some participant negotiators used a deontological decision-making framework, whereas others applied a rationalist or consequentialist framework to make sense of the climate change problem.

Each of the six factors had a unique cognitive structure, one focused on different actor types, the sources of these actors' respective obligations to act on climate change, the shape their action should take, and the structural constraints they had to contend with, the prospects for success, and the emotions associated with these beliefs. Actor types were at the center of all six belief systems, ranging from the individual to groups of countries, to the international community, and even global markets as a vaguely defined collection of firms, industries, processes, norms, and logics, often referred to as market forces. Using an actor type as anchor point, many other elements of a belief system became comprehensible, including threats, interests, and action options. These various elements only attained their meaning in relation to the actor and the actor's relationship to other actors.

This insight has important implications for the strategic change or creation of new belief systems. Cognitive change starts with the player in the game one seeks to influence rather than the problem itself. Creating a strong, positive identity for the player and attaching meaningful norms to this actor in its current setting are key tasks when seeking to affect a belief system. Once the actor identity is established, actor-specific links to the problem can be created, shaping definitions of interests, vulnerabilities, opportunities, norms, and responsibilities. These linkages between the actor, the problem, and potential solutions create the motivations for action on climate change.

The factor comparison provided limited guidance for possible and productive directions of cognitive change. The factor F belief system ("Spotlight on the West") seemed unattractive to state representatives, possibly because of prevailing norms against emotional influences and self-deprecation. A potentially productive endeavor might be to bridge the viewpoints of negotiators having the factor C ("Utilizing the market") belief system and ones having the belief systems of other factors in order to tackle some of the persistent differences in the climate change negotiations, even after Paris. Possible fits include a match between those having the factor C ("Utilizing the market") and those having the factor A ("Multilateralism champions") belief system for global level, state-based action using diplomatic and regulatory means. But economic growth is a major sticking point between these two perspectives. An alternative or complementary approach would be a combination of the factor C ("Utilizing the market") and factor D ("The power of individuals") belief systems, which could help in mobilizing electorates, but different versions would have to be crafted for different cultural groups and voter segments. Integrating factor A, C, D belief systems all together would be a powerful multiscale and multisector approach to the problem.

Apart from identifying potential synergies and links between these six belief systems, it is also important to note fundamental differences between them, in particular, the factor D, E, and F belief systems. Although all three belief systems contain a strong justice-related element, and one could easily group them under a thematic climate justice umbrella, each presents a distinct viewpoint and a different level of integration potential with other belief systems. The factor D ("The power of individuals") belief system is a cosmopolitan view of individual responsibilities and shared humanity—those who have it take each person to task and demand action from others who share this belief system. The factor E ("Climate justice") belief system is the opposite: those who have it see groups of states as the most important subjects of rights and obligations; for them, individual negotiators are merely representatives of groups. Study participants who held this belief system tended to identify with the developing world and framed their positions as demands on the developed countries (i.e., an outgroup). Finally, those having the factor F ("Spotlight on the West") belief system focus on Western responsibility for addressing climate change, an indirect form of self-referential obligation—they call not on individuals but on the individuals' governments to act.

7 A Practitioner's Guide

Seeking to distill some of the more practical lessons of this book, this chapter offers some guidance for practitioners of climate governance. It shows how the diplomats, policy experts, activists, and lobbyists engaged in the UNFCCC process might apply some of the insights presented in previous chapters. Policy makers and stakeholders at the domestic level, who play an important role in the two-level game (see chapter 5) and face challenges very similar to those of the UNFCCC negotiations, might also benefit from what this chapter has to say about translating international obligations into national policies, processes, and institutions.

"Understand, Analyze, and Strategize" presents some reflective-analytic advice, enabling diverse political actors to better understand both their own positions relative to others' and the larger landscape of beliefs that shape the negotiation process. Cognitive-affective maps can help you strengthen your communication skills and make you a more effective political actor. "Meaning Making" explores how you can use CAMs not only as an analytic but also as an interactive tool, one that can support meaning-making processes in climate change negotiations, something particularly helpful when dealing with highly contested or new concepts such as those in the Paris Agreement. "The Big Picture" looks at the practical dimensions of the two-level game. "Strengthen Your Agency" explores ways to bolster both your sense of agency as an individual political actor and the collective agency of your delegation or organization. Finally, "Tackle Science" tackles some challenging questions concerning your understanding and use of scientific information with regard to climate governance. It challenges you to become more familiar and comfortable with some of the trickiest aspects of the climate change problem, such as climate tipping points and long time horizons, and it invites you to learn how to make sense of its many numbers and charts.

Understand, Analyze, and Strategize

First and foremost, the cognitive-affective lens is a potentially powerful analytic tool for political actors to better understand not only their own perspectives on climate change and multilateral cooperation—the various beliefs that form their private belief systems—but also the collective positions of their respective delegations or organizations as well as the positions of other actors and groups.

It is important to distinguish between private beliefs and public positions or arguments. Given the nature of diplomacy, those two cognitive realities might differ significantly in some cases. Usually you only have access to the arguments and statements negotiators voice in public—while their private beliefs remain hidden. Your own mind is the only one you can truly observe and understand at the level of beliefs and convictions. But if you have frequent interactions with other negotiators and if you trust one another, you are probably able to discern many of their true beliefs.

When you analyze belief systems, keep in mind some of the basic principles of cognition. First, ideas and concepts never exist in isolation: they are part of and linked to one another in larger systems of beliefs (see chapter 3). Because our minds have a tendency to maintain the coherence of a belief system, a negotiator will most likely reject proposals that disturb the negotiator's current cognitive-emotional balance without creating a new one. If you would like to change a colleague's view on a particular issue, you will be most successful if your argument is consistent with your colleague's existing beliefs, or if you are able to help the colleague adjust his or her entire network of beliefs to create a new, emotionally coherent belief system. Such systemic change often needs time. Sometimes, you might even have to introduce a new concept, such as nationally determined contributions or global stocktake.

Second, always pay attention to emotions, both good and bad, your own and those of other negotiators. Thought processes and emotions are inseparable, but usually the emotions are not visible. In moments when you can actually observe emotions, as with Yeb Saño's interventions in 2011 and 2012 (see chapter 1), certain ideas or events have probably pushed your colleagues far beyond their normal modes of professional behavior. These moments are not just reminders of the immense scale and human impacts of a changing climate, but also signs that your colleagues might be approaching a breaking point. Respect and try to understand their emotional experiences, which may be very different from your own for many reasons explored in this book, especially in chapter 4. How do your colleagues perceive the threats of climate change? Who are they concerned about? Have they lost something valuable or are they about to?

Every person you interact with holds a belief system with a number of fundamental conceptual building blocks that provide long-term stability for that person's point of view. The three types of concepts to look for include threat perceptions (here, related to climate change), collective identities, and norms of justice (see "The Cognitive Triangle: Threat, Identity, and Justice" in chapter 5).

How different negotiators perceive climate change–related threats varies tremendously. You have probably never experienced or felt these threats in the same way your colleagues from other parts of the world have. But that is neither your fault nor theirs. Even if they wanted to find out what it felt like to be in your shoes, your mental experience is not available to them, given how different your and their life trajectories and identities may be. But there are ways that would allow your colleagues to imagine at least some of your threat perceptions and at least for a time to see the world through your eyes. The most effective way would be to have them visit a place in your home country that is experiencing the impacts of climate change. Observing those impacts firsthand and talking to people who are experiencing them is much more impressive than simply hearing or reading about them.

Even within the negotiations, however, there are a few ways to have your colleagues imagine or see your experiences through your eyes. One such way is storytelling. Relating the experiences of your citizens in the form of stories makes the impacts of climate change specific and concrete rather than generalized and abstract, and thus also makes it easier for your colleagues to emotionally connect with your and your citizens' experiences.

Another, more challenging, way is to develop a shared identity that gives your colleagues reasons to include you (your country, your people) in their circle of in-group members that deserve protection. Our minds' have a natural tendency to separate the world into in- and out-groups, favoring and protecting those in our in-groups—families, tribes, or nations—and treating those outside as less worthy of our protection and care. Depending on the origin of your colleagues, you might not have many identity categories available for such an effort. And the larger the identity group you seek to create, the harder that effort becomes.

But do not assume you know who a colleague is trying to protect. Many climate change negotiators have a number of other identities in addition to their national identities. You might be able to appeal to one of these other identities, for example, that of a parent or grandparent, for the benefit of the negotiations.

It is very likely that many of your negotiator colleagues share a general norm of fairness and care for the vulnerable, but it is also likely there are tremendous differences between their various patterns of moral reasoning. Many negotiators conceive of climate change politics as a fundamental question of justice, which has little or nothing to do with cost-effectiveness. Not acting on climate change and not preventing the human suffering and loss from that inaction are fundamental moral failures that can shake their belief in the moral worth of humanity, the trustworthiness of other negotiators, and the value of the UNFCCC process. But even though all of us are capable of thinking through different moral arguments, none of us consciously chooses a particular pattern of moral reasoning, and especially not the emotional component of moral reasoning. Further, none of us is able to switch at will to what I call a cold consequentialist perspective. Each way of our moral thinking and feeling is a distinct mental response to our perceptions of reality, grounded in the existing, coherent networks of our beliefs.

Threat perceptions, identities, and norms of justice are distinct components of larger belief systems concerning climate change and international cooperation. Chapter 6 described six belief systems shared by a number of UNFCCC negotiators who were participants in a Q study and who often did not even know one another. Each of their belief systems was organized around a specific kind of actor (the state, the international community, the individual) playing an important role in the governance of global climate change. If you can identify the belief systems your colleagues hold, chapter 6 might help you better understand their points of view and engage with them more effectively.

One way to engage in a cognitive self-assessment or to analyze the perspective of another negotiator is to use cognitive-affective maps (CAMs; see "Cognitive-Affective Mapping" in chapter 4) to chart the negotiating position of another person, a delegation, a negotiation bloc or any other group you are interacting with. If a particular colleague is interested, you could consider a meeting in which you use CAMs to map each other's negotiating positions and compare these CAMs to see if you are on the same page. This exercise might reveal some important gaps in your current understanding of each other, opportunities to communicate more clearly, and new possibilities for compromise.

Ultimately, gaining a more detailed understanding of both the cognitive and emotional context of the negotiations will strengthen your communicative, persuasive, and strategic abilities in the UNFCCC process.

Meaning Making

Cognitive-affective maps can support analytic efforts, but they can also be used interactively as a tool to explore and develop shared meanings of complex, contested, or novel concepts in climate change negotiations, such as the new concepts introduced in the Paris Agreement. Some of these concepts—the global adaptation goal, Climate Change Resilient Development (CCRD), and global stocktake, for example—are in need of definition; they are currently empty containers to be filled with meaning—ideally, meaning that the negotiation community shares to a large extent.

Such meaning-making processes take time, but CAMs can facilitate conversations that provide a focus for discussions. You might, for example, set up a three-step workshop with a group of fifteen to twenty negotiators from a variety of negotiation groups including some nonstate actors to make sure that a highly diverse set of perspectives is represented. In the first step, since every new concept in the negotiations has a history and has probably emerged as a compromise between a number of competing versions, you might ask your workshop colleagues to join you in creating CAMs of all the "old" concepts that a specific new concept, the global stocktake (Article 14 PA), for example, is connected to in your minds. See, for example, my CAM of old concepts for the new concept of global stocktake (Article 14 PA) in figure 7.1.

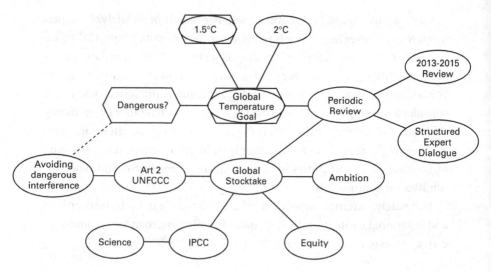

Figure 7.1
CAM of old concepts related to the global stocktake.

Once you and your colleagues have created your own CAMs, you might enter a conversation in which you bring your and their ideas together. Comparing and contrasting individual cognitive-affective maps, the workshop group can develop a collective CAM—a map that you all agree with. Moderating the conversation, you might try to gain as much agreement on the various concepts, their meaning, their relationships, and their emotional valences as possible.

Then, in the second step, you might ask your colleagues to join you in creating your own second CAMs, this time consisting of all kinds of *new* concepts the global stocktake could or should be connected to. Those might be other new ideas in the Paris Agreement or still other ideas you and your colleagues have. For my second CAM of the global stocktake, see figure 7.2.

Again, you and your colleagues might get together to compare, discuss, and, if possible, integrate the various CAMs produced by each of you. Finally, in the third step, you and your colleagues might try to merge the two collective CAMs, exploring how old and new concepts relate to one another. Some old concepts may have become irrelevant or been replaced by new ones. There may be some gaps between the old and new concepts that need to be filled with even newer concepts (e.g., modalities for linking the global stocktake to the periodic review of the long-term goal).

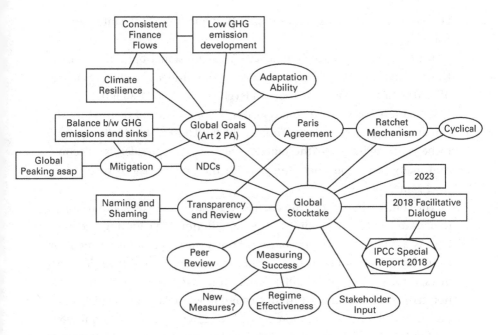

Figure 7.2
CAM of new concepts related to the global stocktake.

Although this process will help each workshop colleague understand the current thought processes of all the others, it can also create some surprising new meanings and ideas that the workshop group can carry into the larger negotiation process. The conversations may be messy, but they will help you avoid misunderstandings and conflicting interpretations in the future.

The Big Picture

Some of the insights offered by this book might enable you to better work in the larger negotiation context. But more important, understanding how the two-level game constantly affects the possibility space of climate change diplomacy will allow you to identify not only barriers other actors cannot overcome, but also potential solutions to specific problems. The two-level game theory concerns the domestic influences or constraints on a country's international politics. These domestic influences, in particular the need to win domestic political support for internationally negotiated agreements, shape negotiation dynamics. Delegations often work in the difficult space between the demands of domestic audiences and those of the international community.

Before the Paris Agreement of 2015, congressional opposition in the United States to any kind of international commitment on climate change offered a case in point for the two-level game. Despite adoption of the agreement, however, the two-level game is not an issue of the past for the climate change negotiations. The case of the United States will continue to be challenging given the outcome of the 2016 presidential election. And many other countries struggle with their own versions of the two-level game.

The complexities of climate change negotiations are vast, and trying to understand what is happening at any point in time is a daunting undertaking. As a political actor, you have to limit your scope of engagement by focusing on a certain topic or a certain set of countries and their demands in a specific negotiation stream. But once you have staked out your negotiation territory, you will want to understand how it relates to the bigger picture and which background forces might be driving the dynamics in your field of interest or responsibility. The analytic framework of the two-level game might help you do that.

First, learn about which domestic political developments of key players are influencing their climate change politics and policy on the ground and try to understand the implications of these developments for each delegation's degree of freedom in the negotiations. Although domestic influences might impose constraints on the negotiations, they might also offer opportunities you can put to best use.

In the case of the United States, the distribution of congressional seats among Democrats and Republicans combined with the Republican stance on climate change policy was a key determinant of the negotiation position of the US delegation ahead of the Paris conference. Even if the delegation wanted to support a more ambitious and more costly agreement, its members—and those of most other delegations—knew that Congress would not ratify any international treaty, at least none with binding commitments. It took a long time to identify a solution space for this particular two-level problem: a Democratic president had to be elected, and that president had to support strong domestic climate change policies. Those conditions were met when Barack Obama took office in 2009. And the international agreement had to be drafted in such a way that it did not require congressional approval but could be ratified by the president him-self; the Paris Agreement managed to meet that condition as well.

Second, if you are a diplomat, make sure you understand and make best use of your own domestic context, especially your domestic stakeholders. To do this, you could create cognitive-affective maps of the negotiation positions of important stakeholders and their interests, comparing their CAMs to a CAM of your own negotiation position. Do their positions help you make the case for certain elements of your own position? Do they contain creative solutions you have not yet considered? Observer organi-zations, from both your own and other countries, can offer a rich source of information that can broaden your understanding of the state of the two-level game.

Or, if you are a representative of an observer organization with a large domestic audience, you could become an active cross-scale actor, building important bridges between the UNFCCC process and domestic politics. Working within the UNFCCC context, you could bring domestic pressure to bear, and not just on your own government. You might help negotiators better understand the domestic conditions in your home country and how that informs the position of your national delegation.

Working within the domestic context, you might be able to create and build political support for action on climate change. Under certain con-ditions, this could shape your government's perceived national interests and open up space within the international negotiations. To that end, how would you affect your audiences' threat perceptions or identities? How would you make a person in California, for example, care about the effects of droughts on ranchers in Texas, or a person in Wisconsin feel for people in Florida who are losing their coastal properties? Understanding how hard it is to imagine the future, how would you help communities imagine the life of their children in a 3°C, 4°C, or 5°C world? What kinds of tools and activities would be necessary for such an effort of imagination? And how would you translate the visions you help create collectively into support not only for climate change policy, but also for change on the ground?

Strengthen Your Agency

"Agency and Hope—Key Ingredients for Changing the World" in chap-ter 5 outlined some of the key dimensions of political agency: a clear idea of goals (both individual and collective) as a focal point for motivations, a good understanding of pathways toward those goals, and a sense of hope that reaching them along the known pathways is possible. Agency also requires access to or at least awareness of the resources for pursuing goals, including knowledge, certain capacities and the power to influence the behavior of other actors (e.g., regulatory power to affect the behavior of firms or political power to affect the negotiation positions of other delegates).

"Agency and Hope" also explored some reasons why negotiators might feel a lack of agency with regard to either their own negotiation group or the UNFCCC community as a whole. Their individual agency might be constrained by their perceptions of the ability and willingness of other delegations to cooperate (e.g., again, the case of the United States): by their lack of power (e.g., many developing countries have no influence over the trajectory of global GHG emissions); or by their not understanding how the shared goal of keeping global warming below 2°C could be attained (pathway thinking). Absent any clear pathways to a 2°C world, the Paris negotiators set their minds on a different goal, which was more clearly within their span of control: a multilateral agreement.

Many negotiators participating in my Q study believed that the set of solutions necessary to address climate change contained changes and activities outside the UNFCCC process and outside the collective control of the negotiation community. Indeed, they experienced no sense of agency when it came to the need for a new global model of development, for example, or the removal of domestic opposition to action on climate change. Their beliefs and feelings in this regard offer two lessons for negotiators.

First, understand and accept the limits of multilateral agency in a multiscale system of climate governance. Some things can be done globally; others not. Recognizing what is within your sphere of influence and what is not will help you avoid frustration and loss of hope. Among those things within your sphere of influence or the collective skill set of the diplomatic community, which are most important to you?

Second, be careful when setting goals. But even though many, perhaps most, of your delegation's immediate goals may be tactical, procedural, or political, do not abandon the pursuit of goals designed to achieve real results on the ground, goals for actually slowing down climate change, dealing with climate change–related impacts, and helping people already harmed by climate change. Which of those goals are most important for your delegation?

When it comes to goal setting at the multilateral scale, it might be worth reconsidering the current set of global goals defined in the Paris Agreement. There is no doubt that the agreement was a big leap forward. Until 2010, there was not even a global temperature goal; indeed, there was no specific global goal for the climate governance regime at all. With the adoption of the agreement, there is not only a staged global temperature target, but also a set of additional goals that address adaptation, resilience, development, and finance. That said, however, none of these additional goals lends itself to measurement or to pathway thinking. Ask yourself and the scientific community what goals *can* be measured and how, and what a pathway toward each goal would look like. How, for example, can the ability of a community or country to adapt be determined, and how can its progress in this ability be measured? What are reasonable milestones and timelines? If pathways and strategies cannot be identified, other goals might be available and more suitable with regard to multilateral agency.

Now that you have accomplished what was previously the most important goal of the negotiations—a new international agreement—how can you shift attention to goals that are linked to the effectiveness of the climate governance regime, in other words, goals that measure global public goods, including climate stability, human, cultural, and state survival, people's health and well-being? Perhaps these goals should include the avoidance of certain climate tipping points—or a new climate governance regime for dealing with climate change–related migration.

If you are a corporate actor, your role in creating, increasing, and maintaining a sense of agency in the UNFCCC process is crucial. Without active engagement of the business and finance communities, none of the goals of global climate governance can be achieved. In the negotiation process, you can provide important information concerning trends and developments in your industry, the potential for and conditions of change, or obstacles that will likely slow the progress of action on climate change. You can help negotiators understand your role in creating and implementing solutions, ranging from emissions controls to adaptation technologies and new financial standards or instruments. Making this kind of information available to negotiators contributes to pathway thinking; it enables them to envision concrete steps toward a different, better future reality, steps that fuel a sense of hope.

Just as important as the role played by climate change negotiators is the one played by nonstate observer organizations. Awareness of their direct contributions toward achieving the global goals as well their engagement of domestic constituencies can be an important source of pathway thinking and hope for negotiators.

Tackle Science

If you have engaged in climate change negotiations for many years, you may feel that your understanding of the problem is sufficient to make you an effective political operative. But even though what you know suffices to participate in the negotiations, which rarely require any detailed scientific insights, a much deeper engagement with science is required to politically achieve the kinds of action that can stave off dangerous climate change and create long-term social benefits beyond the end of this century. Based on the findings presented in "Does Science Matter When You Negotiate Climate Change?" in chapter 5, I encourage all those engaged in the global climate change negotiations to consider novel ways to learn about and understand climate science—to go far beyond scientific in-session presentations, poster sessions, or reading IPCC reports. When looking at charts, numbers, and trends, you need to ask yourself whether you understand what these mean for the future of people not just in your hometown and your country, but everywhere on Earth. You can start with the following four steps.

Engage Your National Scientific Community
The reports of the Intergovernmental Panel on Climate Change aggregate existing scientific findings about climate change and summarize those findings from a global perspective. Serving the international community rather than any individual country, the various Assessment Reports of the IPCC do not—and cannot—fully answer important questions you might have regarding your own country or region. Since the intergovernmental panel does not conduct its own research, but only summarizes what the scientific community has published, it is not the institution to take your questions to. You need instead to engage your national scientific community to

discuss the questions that require the most urgent attention, to increase the probability that your concerns inform the design of future research projects. In addition to climate science, you might want to involve experts from the social sciences, who can help you understand how environmental or technological changes translate into social changes, and how social change can be deliberately created with climate change policy.

Marry Science and Politics
Scientific knowledge is only valuable when you can connect it to your political analyses, goals, and strategies. On the one hand, it can inform you why, for example, drought or flood conditions in a certain region are changing, why a certain disease might be on the rise, how the intricate dynamics of natural systems have been altered due to a range of human behaviors that generate GHG emissions and land-use changes, and what changes we might expect in the future. But, on the other hand, scientific knowledge can also inform you about solutions to the climate change problem. What is the scientific foundation of your delegation's goals? Why is 2°C, or perhaps 1.5°C, a reasonable temperature target? Do you know what the likely effects of 2°C warming compared to those of 3°C warming might be on your hometown, on your country's agricultural productivity, on the shape of its coastline? Do you know the adaptation options and their respective costs for your communities in a 2°C compared to a 3°C world? Can scientific knowledge help you understand potential limits to adaptation and what to do in those circumstances? How should this knowledge inform the review process under the Paris Agreement, the work of the Adaptation Committee, or the Warsaw International Mechanism on Loss and Damage?

Pay Attention to Climate Tipping Points
Because they have major disruptive potential and could easily affect your chances to achieve your political goals, climate tipping points deserve your attention. Do not assume that climate change always or primarily happens in linear, gradual ways—a little more each year. It can be dramatic and sudden, potentially overwhelming the response capacity of the affected systems, which can be radically altered, with profound ripple effects in unexpected places. A tipping point process very distant from your home country, for example, the Arctic, Greenland, or the South Pacific, can affect

you and your citizens in unexpected ways. Avoiding specific climate tipping points might therefore become an important component of your political goals. Indeed, awareness of climate tipping points has major implications for global climate governance and should inform future global goal setting and institutional design.

Time and the Future

Many of your responsibilities require that you pay attention to the immediate: the events of the day, a water shortage this season perhaps, or a change in government coming up next year and the mood of the electorate during the election season. But climate change requires you to take a step back and grapple with the long term. Find out what the long term means—how far in the future your decisions will actually have an effect. Explore your own ability to imagine the future state of the world, especially that of your hometown or home country. What does climate science tell you about life in 2047 or 2062? Or even 2116? Is an 8.5°C world even imaginable in some meaningful way? How warm would such a world be on a summer afternoon? Could you spend a day outdoors? Under what conditions? What plants, trees, and crops would survive? What would farming look like and where would it take place?

These really hard questions have no straightforward answers. The scientific community might be able to offer some clues, but the challenging cognitive-emotional task of imagining possible climate futures for humans, plants, and animals is one that you yourself must undertake: your mind has to turn scientific knowledge into meaningful images of and beliefs about the future. And to do this, you need to better understand how to boost your mental capacities.

There are specific techniques that can help you get better at imagining the future, such as scenario planning and participatory visioning workshops (Sheppard et al. 2011; Wiek and Iwaniec 2013), although these techniques have not yet been deployed in the UNFCCC context to address challenges at the science-policy interface. If you are interested in engaging in more meaningful future thinking, reach out to the scientific community, both at home and abroad, to get support.

Beyond the use of well-established tools, developing your future-thinking capacities as a UNFCCC negotiator might require additional, and perhaps less conventional, means. For example, the growing artistic and literary engagement with climate change might offer opportunities for UNFCCC negotiators to move beyond the verbal engagement of climate science to more visceral-emotional and graphic forms of engagement and social learning. Narratives and storytelling, including the literary genres of climate change and speculative fiction, offer unexplored potential to advance science communication that engages the entire suite of our cognitive-emotional sense-making abilities (Milkoreit 2015, 2016).

Summary

My key suggestions for putting *Mindmade Politics* to best use in your daily practice of climate change diplomacy, activism, and decision making include the following:

• Actively deepen your understanding of your own negotiation position and those of your colleagues using a cognitive-affective lens.

• If you seek to change a colleague's mind, pay attention to the systemic nature of that colleague's beliefs and the need of the colleague to maintain cognitive and emotional coherence. Three types of conceptual clusters form the building blocks of belief systems in the climate change negotiations: threat perceptions, identities, and justice.

• Use cognitive-affective mapping to develop shared understandings of contested or novel concepts.

• Use the idea of the two-level game in the larger negotiation context to learn about domestic constraints and opportunities of important actors.

• Strengthen your own agency with a deliberate focus on possible, ambitious, and measurable global goals that lend themselves to pathway thinking.

• Engage with climate science, in particular, with the concept of climate tipping points and its implications for global climate governance. Train your mind to expand its naturally short time horizon, to grapple with the long time scales of climate change. And dare to think about the future—all kinds of futures. Live up to your responsibility to better understand what kinds of futures are possible and what the international community can do to create the most desirable future possible.

8 Thoughts That Make the Future

Mindmade Politics has sought to answer one fundamental question: what is the role of the mind in international climate change negotiations? Exploring whether and how cognitive processes shape global climate change politics is not just interesting from a scholarly perspective, it may also be vital for progress on meeting one of the most fundamental challenges facing humanity in the twenty-first century. Having discovered a whole suite of intriguing insights about cognition, emotion, belief systems, and climate change diplomacy, I find myself at the beginning rather than the end of a research program on political cognition.

Despite the huge complexity of the issues covered by the global climate change negotiations and the diversity of the individuals involved in this research, a number of interesting insights have come to the surface and invite further exploration. With the hindsight of the Paris Agreement, I was able to reflect on the early stages of the negotiations that culminated in the adoption of the agreement in 2015. The data collected in 2012 opened a window into the minds of negotiators at a point in time when it was not yet clear whether there would be a new agreement, and which ideas it would contain. Looking back, I was able to connect some of the concepts discussed in 2012 with the results of the Paris Conference of the Parties (COP 21) in 2015, distinguishing cognitive facilitators of and obstacles to success.

Engaging a subset of carefully selected UNFCCC negotiators from the distinct communities of diplomats and observer organization representatives, I identified cognitive elements—distinct concepts, beliefs, and goals—that shaped their individual points of view (chapter 5), but also a set of larger belief systems that existed among groups of negotiators (chapter 6). Some of the mental patterns of study participants focused on national politics and the importance of domestic factors. The majority of their thoughts, however, concerned the international dimensions of climate governance. This might be the greatest power of a cognitive approach: the ability to bridge and connect the multiple scales of climate change politics, from the individual to the global.

Below I briefly summarize the most important insights of *Mindmade Politics*. I hope they will enrich existing debates in political science and provide cognitive scientists a new application of their growing arsenal of conceptual tools. Some of these observations may also provide some initial guidance to diplomats, policy makers, climate scientists, and knowledge brokers regarding the communication strategies needed to help create the conditions for effective multilateral cooperation.

Cognition in Global Climate Change Politics

In the course of this book, I have developed three connected sets of insights. The first set concerns the nature of cognition and the contribution of a cognitive approach to the study of international relations. The second set addresses the characteristics of belief systems of political actors engaged in global climate change negotiations—a case study of political cognition in global environmental governance. The third set identifies specific cognitive challenges these actors experienced when dealing with the special problem characteristics of climate change. Together, these insights provide a deeper understanding of the nature of agency and related beliefs in the UNFCCC process.

The Nature of Cognition

A cognitive approach to international relations scholarship highlights three aspects of political life that other theoretical perspectives usually ignore (see box 8.1): (1) the cognitive patterns that create meaning, purpose, and motivation for political behavior (i.e., the mental sources of political agency); (2) the relationship between individual and collective beliefs and agency in political life; and (3) the role of emotion in political thought and action.

Most important, all concepts and ideas in our minds are part of larger systems of beliefs. Discrete concepts are linked in a unique way that maximizes the emotional coherence of the larger systems—in this cognitive

---

Box 8.1

Elements of a Cognitive Approach

1. Cognition is a systems phenomenon
2. Mental representations linked to physical or social reality
3. Person-group relationship (collective cognition)
4. The role of emotions

---

architecture of meaning, things have to make sense, both logically and emotionally. Given the contribution of each concept and its relationships to the system-level property of coherence, individual concepts and beliefs are often hard to change. Removing or replacing them disrupts the architecture's "cognitive balance." Cognitive change requires its reconfiguration in an emotionally coherent manner.

Individuals develop emotionally coherent networks of mental representations—belief systems—in the course of their lives by interacting with other individuals (i.e., the social environment) and the physical-material environment. Meaning making is a relational process that uses language to link physical realities with mental representations. The same mechanisms apply to socially constructed abstract entities such as democracy, identity, or justice, where meaning making relies on symbols, behaviors, conceptual definitions, and other abstract representations. Belief systems are cognitive structures that enable and constrain political thought and decision making.

A central issue for understanding cognition in political processes is the individual-group problem: how do groups of individuals come to share a set of beliefs? How do groups change their "collective mind"? Since collective thoughts have no foundation in a group brain, collective cognition is best understood as a multilevel mechanism that links cognitive processes at the individual level with social communication processes at the group level. Individuals think of themselves as members of groups. They have thoughts and feelings about their groups as group members and mental representations of the groups themselves as entities. Group members share these ideas with one another; their group-related beliefs converge over time. Collective cognition is a recursive interaction process between individual minds, on the one hand, and social networks of the minds of multiple people, on the other.

Finally, when researchers view political decision making through a cognitive lens, they are forced to acknowledge the role played by emotion, broadening the set of relevant, interacting variables and challenging the methodological toolbox of international relations scholarship.

The Nature of Political Belief Systems

Linking cognitive theory to international relations, I have explored the mental representations of climate change and multilateral cooperation in the minds of individual negotiators. I have offered evidence that both the rational-choice and the constuctivist schools of thought make important contributions to the analysis of global climate change politics. Indeed, though both are necessary, even jointly, they cannot explain all observable political dynamics (see box 8.2).

Using the cognitive assumptions of both rational-choice and constructivist approaches to develop an analytic framework, my analysis identified three types of mental representations that existed in the belief systems of all study participants: threat perceptions related to climate change, group identity concepts, and norms of justice. These three conceptual clusters formed a cognitive triangle, in which all three corners influenced one another. Different constellations of the concepts at each corner of the triangle generated different kinds of belief systems that could be observed in the climate change negotiations.

Distinct patterns of moral reasoning helped distinguish between two major types of belief systems. Emotionally driven differences in climate change threat perceptions combined with different conceptions of a study participant's in-group (collective identity) determined whether a particular belief system of that participant integrated deontological or consequentialist norms of justice. Participants who expected threats above a severity threshold of human suffering for their in-groups tended to apply a "hot"

Box 8.2

Key Features of Political Belief Systems (The Cognitive Triangle)

1. Diverse threat perceptions and associated emotions
2. Collective identities, especially place-identity, and associated emotions
3. Patterns of moral reasoning and associated emotions
4. Hot deontology and cold consequentialism

deontological framework of reasoning that was often associated with strong emotions. Those who were concerned about threats below the severity threshold, mainly economic and material losses, tended to apply a considerably less emotional ("cold") consequentialist framework of reasoning. The quality of the expected threats for participants' identity groups was the decisive factor in differentiating their belief systems.

These cognitive patterns split the participant negotiators into two major groups: those concerned about the costs associated with climate change impacts and those concerned about the costs of policy and action on climate change. These two groups did not readily coincide with the familiar categories of developed and developing country representatives. Many negotiators from the developed world shared concerns about climate change impacts that triggered normative beliefs about cooperation—they believed that it was simply morally right to act collectively on climate change.

Most study participants associated themselves with multiple identity groups. The state or the organization they represented was central and often their most important collective identity for a political actor. But additional groups could complement and even compete with a national or organizational identity. Some of the participants' belief systems focused on individuals, others on groups of states, and some even on planet Earth as an all-encompassing identity group that contained humanity and all life on Earth.

The importance of group identities cannot be overemphasized: they determined which threats individual participants were concerned about. The combination of threat and identity shaped the participants' entire belief systems, including their willingness to commit their respective groups to cooperation in a multilateral forum. Starkly contrasting identity conceptions, for example, national interests versus humanity sharing responsibility for one planetary resource—led to starkly contrasting preferences for the timing and scale of multilateral action. More inclusive identity groups made it more likely that participants expected their groups to be threatened by severe climate change impacts. Consequently, the likelihood of favoring multilateral cooperation and immediate action increased with the growing inclusiveness of the participants' in-groups.

Many study participants mentioned identity groups not encompassed by the current rules of the international system, such as "the poor" or "humanity." International law does not accord these groups any rights or responsibilities. This observation reveals an important incongruity between the existing conceptions of climate change as a global governance challenge and the available multilateral tools for addressing the problem. Some participant negotiators believed, and I concur with them, that solving climate change requires the development of collective identities commensurate with the global scientific definition of the climate change problem, and consequently the involvement of groups and categories currently not acknowledged in international politics.

Place identity was particularly important for the formation of participants' belief systems and their negotiation positions concerning global climate governance. Place identities in their great diversity proved to be a good proxy for threat perceptions, for instance, the anticipation of physical changes in locations linked to a collective identity. Place identity or sense of place is an emotional phenomenon: anticipating or experiencing the loss of these places causes dread, anxiety, fear, anger, and sadness. Further, place identity changes the relevance and emotional implications of climate change–related extreme weather events, which become tangible, visible traumas to spaces groups care about in ways that most of the current policy discourse surrounding extreme events does not recognize. Connecting physical-material structures like territory or land with ideational structures like identity or culture, the concept of place identity not only adds explanatory value but also offers opportunities to bridge realism and constructivism.

To the extent that participants' belief systems contained concepts about norms of justice in global climate governance, they were far less contentious than past negotiation positions and political dynamics would have suggested. The concepts of equity and historical responsibility were not as prominent in the interview data as they were in the negotiation process leading up to Copenhagen as well as Paris. Very few study participants considered equity and historical responsibility important and appropriate for future agreements. More important were notions of fairness, caring for the poor, and the idea that all of us should take responsibility to the degree we are able. This difference between private beliefs and negotiation positions signaled a normative trend away from the contentious politics that had prevailed prior to 2012, and indicated that the equity debate might in some instances be a cover for other interests or concerns. Norms were important, but formal negotiation positions and interventions did not reflect parties' substantive concerns and normative beliefs very well.

The Paris Agreement can be read as a consequence of this discrepancy between beliefs and positions: although it contains many instances of differentiation between the more and less vulnerable as well as the more and less capable, it makes no reference to historical responsibility. Most important, however, it establishes the same kinds of obligations for all parties, regardless of their development status. This is a massive shift away from the model of Kyoto Protocol, which created obligations for developed countries, but not for developing ones.

These observations speak directly to existing theories in international relations scholarship and are to a large extent enabled by those theories. But *Mindmade Politics* also explored a number of issues that do not fit the existing theoretical categories as easily.

Special Problem Characteristics of Climate Change

Many study participants displayed at best a shallow understanding of some of the characteristics of climate change that make it the daunting political problem that it is. The governance implications of these characteristics (see box 8.3) have so far not been explored, let alone addressed in the negotiations. One problem characteristic stands out: climate tipping points.

When starting out, I suspected that many participant negotiators would be struggling with a loss of hope in response to the pervasiveness of climate change. I was wrong. Participants acknowledged the pervasive nature and associated uncertainties of climate change, but were generally not discouraged by these observations. A specific subset of individuals, especially negotiators from small-island states, indicated that they had occasionally and temporarily lost hope in the UNFCCC process, but not because of the complexities of climate change. It was because high-emitting countries declined

---

Box 8.3

Cognitive Challenges Dealing with Special Problem Characteristics

1. The productivity and necessity of hope
2. Links between poorly defined global goals and perceived lack of agency
3. Limited imagination regarding the long-term future
4. Insufficient attention to climate tipping points

to act on climate change despite the devastating consequences it would have on small and highly vulnerable countries.

More generally, hope prevailed among participant negotiators although there were very few rational reasons in 2012 to be hopeful about the ability of the international community to deal with the climate change problem effectively. The search for hope and avoidance of negative emotions, such as anxiety over lack of control, led many participant negotiators to abandon the 2°C temperature target and to replace it with a more achievable goal for themselves: a negotiated agreement. Maintaining hope with respect to this more moderate and qualitatively very different goal allowed the political process to continue, which facilitated the Paris Agreement. Although the agreement provided a new source of optimism and gave a major morale boost to environmental diplomacy, it is unclear whether it will actually make a difference in people's lives. Does it offer real hope for curbing, containing, and managing the impacts of climate change?

One of my initial hypotheses concerned the limited observability of climate change. I expected that the participants' general lack of experience of climate change impacts might limit their motivation for collective action and hence inhibit cooperation. But, for most negotiators from the developing and emerging economies, climate change was already an observable phenomenon in 2012. Many had either had a personal experience of climate change or referred to reports from the citizens they represented. In the developed countries, a similar trend has been developing in the aftermath of Hurricane Sandy, the occurrence of the Polar Vortex, frequent droughts, and floods and wildfires in the United States, Europe, and Australia, although participant negotiators from these countries rarely had a direct, personal experience related to climate change and did not expect to be personally affected in their lifetimes.

Nevertheless, based on my study data, the limited observability of climate change impacts does affect decision making in the UNFCCC. In conjunction with the long time scales of climate change and system response lags, limited observability impairs three crucial governance tasks: (1) imagining the long-term future and implications of governance failure; (2) developing action timelines; and (3) setting and acting on goals and targets for action on climate change.

1. The majority of study participants were unable to imagine a nonlinear, qualitatively different future, in which some of the more extreme

possibilities of climate change might have materialized. Most participants had never encountered this cognitive challenge before. Their
lack of imagination, though not surprising—climate change presents
an altogether novel challenge to our human minds—has an important
effect on present decision making in the UNFCCC. If negotiators do
not actively consider a large spectrum of possible futures, for whatever reason, and if they are not aware of the causal influence they collectively have over different possible futures, they cannot make fully
informed decisions. In this sense, cognitive myopia poses serious challenges for the negotiation process, which go far beyond "rational" discounting of the future. The Paris Agreement established the maximum
time horizon currently relevant for UNFCCC negotiators: the Nationally Determined Contribution (NDC) review process up to 2030.

2. It was hard and often impossible for study participants to identify clear
    timelines for impacts of or action on climate change, or to link particu-
    lar actions with particular outcomes on a time scale. Only a subset of
    negotiators from highly vulnerable countries focused on timelines pre-
    scribed by the scientific community, including a peak year for global
    emissions and target years for certain levels of global emission reduc-
    tions (e.g., 80 percent by 2050). For many other study participants,
    time was not an important variable in the negotiations or in designing
    governance instruments.

3. Limited observability and challenges when dealing with the long-term
    dimensions of climate change also impacted participants' thoughts
    about climate governance goals. They lacked a clear definition and
    shared understanding of the goal or goals of climate governance, and
    they focused on limiting global warming to 2°C as the central goal of
    the climate governance regime, although many had started to replace
    the 2°C goal with the diplomatic goal of reaching a multilateral agree-
    ment. The main reasons for privately abandoning the 2°C temperature
    goal included concerns about the slowness of the political process and
    doubts about governance regime effectiveness. Participants did not
    know how, in practical terms, to reach the temperature goal, which
    undermined their sense of hope and agency. The move from what was
    perceived to be a scientifically determined goal (2°C) to a diplomatic
    one strengthened their sense of hope and agency; it gave negotiators a
    stronger sense of control over the process, placing the desired outcome

squarely within their personal and collective skill set. At its heart, this mental exchange of goals demonstrated the disconnect between individuals' assessments of what was politically feasible (a multilateral agreement) and what was necessary for environmental regime effectiveness (the temperature goal).

Study participants tended to think about global average temperature increase in a linear fashion: a 3°C world would be a little worse than a 2°C one, and a 4°C world would be better than a 6°C one. This linear conceptualization of climate change was consistent with their personal experience and was linked to their limited ability to imagine climate tipping points or a qualitatively different long-term future.

In a surprising twist, the Paris Agreement not only confirmed the 2°C goal that many participants had doubted in 2012, but even included the more ambitious, and at this point fairly unrealistic, goal of pursuing "efforts to limit the temperature increase to 1.5°C." I doubt that many negotiators believed in the achievability 1.5°C when the Paris Agreement was adopted, but they most likely accepted the language for the sake of political compromise or because of its symbolic value, welcoming the inclusion of 1.5°C for tactical reasons: it had short-term benefits (reaching an agreement) but no long-term political costs. As long as countries could argue that, collectively, the parties to the agreement pursued "efforts" toward the 1.5°C goal, no single country could be held responsible for not reaching the goal.

The Paris Agreement also added a set of new global goals that reflect a broadening of the purpose of global climate governance. But because many of those goals will be hard to define and harder still to measure, they might further limit negotiators' sense of agency.

Of even greater concern than goal selection was the general lack of attention to the possibility of climate tipping points among participant negotiators. Rather than dismissing the importance of tipping points for the governance process (e.g., because of its weak scientific basis), many negotiators were not familiar with the concept at all and did not incorporate tipping points into their negotiation positions or beliefs about climate governance. Indeed, no study participant suggested that preventing climate tipping points might be a useful climate governance objective or that tipping points could disrupt pursuit of the goals of the UNFCCC. Despite the quickly growing use of the concept in various scholarly communities, based on my study data, climate tipping points did not receive much attention in the diplomatic and NGO communities in 2012. The Paris Agreement has not changed this situation.

Other Cognitive Challenges

My analysis of study participants' belief systems further revealed that the United States played a special role in the climate change negotiations before Paris. There was an unusually high level of awareness of the domestic challenges in the United States that had been impeding the introduction of effective climate change policies for many years. Many participant negotiators harbored doubts that the U.S. delegation was able to commit the country to a meaningful agreement; these doubts placed strong constraints on their beliefs about their collective agency in the UNFCCC process. This unique two-level game setting raised interesting questions for regime theory. It suggested that, in the case of global climate governance, regime creation without the hegemon's participation was simply not possible.

Bilateral collaboration between the United States and China and a careful crafting of the Paris Agreement have resolved this United States–specific problem temporarily. But such collaboration points to two additional problems for global climate governance. First, there is a large set of partial solutions to the climate change problem outside the scope and reach of the UNFCCC negotiations. Although there can be no reasonable doubt that such partial solutions are complementary to the work of the UNFCCC negotiators, when these negotiators identify them as necessary preconditions for an effective international agreement, they become serious obstacles to multilateral action. Preconditions such as the adoption of a new development model not centered on growth, the mobilization of domestic constituencies by changing their values and preferences, or changing the minds of political elites not involved in climate change negotiations cannot be met by UNFCCC diplomats and indeed depend on developments beyond their control. Considering these preconditions to be necessary thus severely constrains the diplomats' agency and meaningful progress in the UNFCCC negotiations.

Second, framing US domestic politics as a serious constraint for multilateral treaty making raises the broader question about the relationship between democratic government and global governance in addressing global problems. In the case of climate change, past experiences with the

United States, Canada, Australia, New Zealand or, more recently, Poland highlight that democratic opposition to engagement with certain problems can stall global progress, affecting the fate of people far beyond the borders of the obstructing democratic countries. Domestic developments can cripple perceptions of multilateral agency, and democracy can become an obstacle to global governance. This obstructive role is likely to grow more important with an increasing number of truly global challenges requiring multilateral cooperative solutions.

Six Belief Systems

Stepping back from the specific cognitive elements and patterns identified with the help of cognitive-affective maps, my Q analysis offered a close look at some of the existing belief systems within the larger community of climate change negotiators. Whereas the CAM analysis focused on participants' different conceptions of the kinds of threats that mattered or the moral nature of climate change, the results of the Q analysis pertain mainly to who the participants believed were the most important actors, what they should do, and why.

Each belief system links various cognitive elements to form a coherent whole. Due to the limited set of Q study participants, the six belief systems described in chapter 6 likely do not represent the full spectrum of the UNFCCC negotiators' views on climate change and multilateral cooperation. But they offer a first in-depth look into the complex cognitive dynamics driving the global climate change negotiations.

Almost all study participants shared a number of fundamental beliefs, such as the importance and reality of climate change, trust in science, and the collective ability of the international community to solve the problem. Opinions diverged sharply on a small number of issues, including whether all or just a few countries should be at the negotiation table, the distribution of climate governance tasks between governments and markets, and, ultimately, the importance and desirability of economic growth.

The key element distinguishing the six belief systems briefly described below was actor type: each belief system had a different type of political actor at its core; this has implications for problem definitions, proposed solutions, and political strategies with regard to climate change. The actor type provides the anchor for creating an emotionally coherent system of beliefs.

Multilateralism Champions Study participants holding this belief system focused on the collective agency of all states that were parties to the UNFCCC, and considered them to be bound by ethical norms rather than national interests. They paid little attention to domestic processes or other factors that might constrain global efforts and negotiation positions. They viewed the key challenge as overcoming the North–South divide in order to create a cooperative multilateral solution. They considered growth a problem rather than a solution to the climate change challenge.

UN Skeptics Strongly preferring a multilateral club approach to global climate governance, participants having this belief system wished to engage the major emitters only and to avoid the cumbersome and, as they saw it, ineffective UN process. Grounded in facts and scientific knowledge, their long-term perspective gave rise to strong concerns about future climate change impacts, including the possibility of climate tipping points. They believed that responsibility to avoid climate change–related harm lay with elected officials in domestic political contexts, where most of the policy work needed to be done.

Utilizing the Market Study participants having this belief system linked the need for multilateral cooperation on mitigation among all states with the use of market-based instruments, such as a carbon-based price signal, to bring about needed socioeconomic changes. They believed that climate governance was a two-level game, that stronger voter support was needed to overcome systemic vested interests, and that action on climate change was a necessity and a collective-action problem, not a moral issue.

The Power of Individuals Being skeptical of neoliberal policy frameworks, participants with this belief system had confidence in the power of individuals to create change, ranging from individual value change affecting consumption patterns to change in local and community engagement and in the exercise of democratic rights to vote and to mobilize other people for change. Their viewpoint was grounded in a strong human identity—shared with people around the world rather than with fellow citizens only—and in a moral framework of obligations binding rich and poor as well as present and future generations.

Climate Justice   Study participants who shared this belief system assigned responsibility for fixing the climate change problem to the developed world for two reasons: historical emissions and present resource wealth, the latter being the basis for adaptive capacity. They believed that adaptation and support for the developing countries were key components of a just climate change solution, which had to prioritize the needs of the vulnerable developing countries, and that scientific information was crucial to motivate a North–South narrative about past exploitation and present moral obligations to alleviate suffering from climate change–related impacts.

Spotlight on the West   Participants having this sixth and last belief system focused on the responsibility of the developed world, too. They saw the need for value change in Western societies as an ethical responsibility of the present generation to future ones. Taking themselves and their own societies and governments to task, individuals with this perspective advocated for more sustainable, more socially connected, and less consumption-oriented lifestyles. They also expressed great disappointment and shame about the continuous failure of their own societies to address climate change.

Most belief systems shared a surprisingly similar emotional profile in the sense that participants who hold very different points of view have a very similar emotional experience. The outliers were "Utilizing the market" and "Spotlight on the West" participants, who had opposing responses to negative self-referential emotions like shame. Those who believed in the power of markets strongly denied the importance of these negative emotions; those who focused on the responsibility of their own Western societies embraced them. These two belief systems stood out in other ways, too. "Utilizing the market" participants were the only ones who did not include explicit norms of justice in their belief system, and "Spotlight on the West" participants seemed to be almost exclusively NGO representatives.

A comparative look at "The power of individuals," "Climate justice," and "Spotlight on the West" belief systems reveals a number of striking similarities but also important differences between these perspectives. All three contain a strong justice-related element and could easily be grouped under a thematic climate justice umbrella. But each presents a distinct viewpoint and different levels of integration potential with other belief systems. "The power of individuals" belief system was characterized by a

cosmopolitan sense of individual responsibilities and shared humanity—
participants holding this viewpoint believed they had to act on it. "Climate
justice" presents the opposite viewpoint in many ways. Participants who
held it referred to groups of states as the most important subjects of rights
and obligations and treated individual negotiators merely as representatives of states. They tended to identify with the developing countries and
framed their position as demands on the developed countries (i.e., an outgroup). And participants with "Spotlight on the West" viewpoint believed
that Western governments and societies, not individuals, were responsible
for and called to action on climate change.

Interestingly, most study participants held parts of two or three different belief systems. This pattern of belief system combination indicates that
there is large potential for bridging these partly overlapping views, possibly integrating them, and building broader or different—hopefully more
productive—negotiation groups.

But the possibility for creating belief change does not imply an argument in favor of belief convergence. Given the complexity of the climate
change issue and the diversity of views about climate governance, belief
convergence would be both unreasonable and unnecessary. It appears more
promising to seek "clumsy solutions" to the "wicked problem" of climate
change, which would allow different actors to pursue similar climate governance goals for very different reasons (Verweij et al. 2006).

Finally, it is worth noting that most of the six belief systems examined
were shared by participants from very different countries, regions, and
negotiation groups, which suggests that their viewpoints were at least to
some degree independent of material variables like national emission levels
(i.e., cost of mitigation) and vulnerability. The fact that the distribution
of the participants' beliefs does not match the existing divisions between
negotiation groups is also encouraging for the future of the negotiations.

Negotiation Process, Political Strategy, and Science Communication

Beyond revealing the content and structure of political belief systems, *Mindmade Politics* has offered policy-relevant insights in three distinct spheres:
(1) the multilateral negotiation process; (2) the development of negotiation
strategies and positions at the national level and by nonstate actors; and
(3) planning and communication efforts of the climate science community.

The Multilateral Negotiation Process

The previous chapters have revealed a range of private beliefs and mental patterns that feed the negotiation process but often remain invisible. The cognitive processes of participants engaged in global climate change negotiations displayed a set of remarkably similar structural features (e.g., the cognitive triangle, actor focus) and consistent mental patterns (e.g., concerning science and climate governance goals), but also a rich diversity in terms of the content of their belief systems. Insights into the motivating beliefs and values driving different actors can help climate change negotiators gain a better understanding of the negotiation dynamics they are part of. A deeper understanding of the nature of negotiation positions, especially the resistance to change due to the embeddedness of concepts and beliefs in larger belief systems, can inform negotiation strategies and behavior. Further, acknowledging that the sources of contention might sometimes be rooted in the different makeup of individuals' minds and their different cognitive-affective experiences of reality might help improve communication and trust among negotiators.

Important issues that divided participants from the parties to UNFCCC negotiations in 2012, but that were seldom openly discussed, included the possibility of a multilateral club approach to complement or replace the UNFCCC process, and the multilateral paralysis caused by domestic politics in the United States (i.e., the importance of the two-level game). With the successful adoption of the Paris Agreement and the momentum it created, both of these issues will move into background for the time being.

My study data also revealed that the equity debate could become far less contentious if moved away from the narrow idea of historical responsibility to a broader notion of responsibility shared by all countries, based on "respective capabilities" and a concern for the poor. The cognitive seeds for such a change existed back in 2012, and have grown in the process of negotiating the Paris Agreement. But questions of equity will continue to shape the negotiations in the future, while the normative expectations of countries like China or India remain in flux.

The election of Donald Trump as the next President of the United States could undermine the progress that has been made in these two areas. The new administration has the potential to recreate two-level game challenges and consequently elevate equity concerns if the U.S. is perceived to shirk its global responsibilities.

With a view to the future, the analytic tools presented and applied in the previous chapters could also be deployed in the process of implementing the Paris Agreement. The need to fill the agreement's key terms with meaning and to develop shared understandings among negotiators with very different belief systems and negotiation positions will accompany the negotiation process for years to come. Cognitive-affective mapping can support the process of developing, disputing, and agreeing on definitions of concepts and their relationships with one another.

Developing Negotiation Strategies

Taking the cognitive status quo as a point of departure, delegations, national policy makers, and nonstate actor groups can adjust their own political and negotiation strategies to tailor their messages, offers, and demands to different audiences in the negotiation process. Recognizing the links between cognitive structure and content, the need for maintaining coherence, and the importance of emotions, they can improve their ability to speak to the concerns and values of other negotiation parties, and possibly even to change existing belief systems. Such efforts at change can be internal to in-groups or external, focused on out-groups. They can take place at the multilateral scale or at other scales, all the way down to local communities or individuals.

The Science-Policy Interface

My analysis has also provided clues regarding current communication failures at the science-policy interface, challenging the scientific community and the UNFCCC negotiators to rethink their approaches to science communication. Two topics deserve particular attention: climate tipping points and the ability to imagine the distant future. Without deliberate efforts to increase negotiators' capacity to deal with the cognitive challenges presented by climate change, through training in systems and future thinking, for example, regime-building efforts will remain constrained and most likely ineffective.

These cognitive challenges raise two questions. Why is existing scientific knowledge not being heard or used? And what types of information or forms of communication are needed to help policy makers grapple with the special problem characteristics of climate change? Although these questions go far beyond the scope of this book, I have suggested in chapter 7 that new forms of science communication might be needed, including a role for art, visual representations, and stories that engage not only the cognitive but also the emotional and experiential faculties of climate change negotiators.

More generally, the science-policy disconnect revealed through the analysis of the beliefs of study participant negotiators indicates that it might be useful, even necessary, to rethink the current interaction rules and modes of communication between the Intergovernmental Panel on Climate Change and diplomats in the UNFCCC process. If current forms of presenting scientific information—such as the research dialogue of the Subsidiary Body for Scientific and Technological Advice (SBSTA) or the UNFCCC's Structured Expert Dialogue for the Periodic Review—are insufficient to create beliefs that reflect relevant scientific knowledge, additional avenues of communication should be considered.

Reflection

The aim of this research project was twofold: to map the existing cognitive landscape in the global climate change negotiations and to establish whether and how mental processes affect political dynamics. I have argued that the global climate change negotiations are being stifled by a battle of belief systems that differ fundamentally in their conceptual—especially normative—and emotional makeup. Many negotiators have major difficulties dealing with the special characteristics of climate change and integrating scientific knowledge into political positions, which reduces their willingness to cooperate and their sense of urgency, weakens their sense of agency, and negatively affects the definition of climate governance goals and targets. Although the Paris Agreement demonstrates the ability of the international community to develop compromise solutions and even to innovate in the face of long-standing adversity, failure, and contention, the new agreement has not resolved the problems discussed.

It is possible that, at this stage of our evolution, we humans do not have the cognitive ability to address the novel characteristics of climate change. Given our lack of previous opportunities for learning about the nature of this problem, our minds may simply not have the appropriate tools—conceptual frameworks and emotional responses—to generate beliefs that can produce proactive, collective solutions to climate change. Existing philosophical or moral frameworks may be insufficiently developed to address the question of responsibility for humanity rather than individuals, nations, or states. The two strongest candidate frameworks, utilitarianism and cosmopolitanism, do not speak to issues of intergenerational justice or the value of planetary systems essential to human life and well-being. More generally, discussions of intergenerational justice at a global scale are still in their infancy (Gardiner 2011) and remain abstract and ungrounded in experience. Economic frameworks have been developed to understand market dynamics over several years, but not over decades, let alone centuries. State institutions take a similarly short-term perspective with typically four- or five-year election cycles. Our collective identities have grown from tribes and clans to nations and states, but humanity as our ultimate in-group or collective identity seems far beyond our reach. With few exceptions, study participant negotiators' imagination and empathy did not reach far enough beyond state borders or far enough into the future to facilitate timely collective action that could slow climate change.

Recognizing these cognitive constraints on global climate change governance is the first step toward overcoming them. Understanding the sources of current differences in negotiation positions based on belief systems could improve mutual understanding among climate change negotiators, increasing much needed trust and mutual respect in the negotiation process. The lack of attention to climate tipping points calls for changes in the science-policy interface, including different means of communicating scientific concepts and engaging negotiators in discussions about their implications for climate governance. Participant negotiators' general inability to imagine the distant future suggests that innovative ways are needed to represent, visualize, and explore different possible futures both conceptually and emotionally so that negotiators can better understand how their collective actions make some of these possible futures more or less likely.

If the mind is an obstacle to cooperation, changing minds becomes a political necessity.

Research Outlook: An Expanding Field of Cognitive International Relations?

With *Mindmade Politics*, I have begun to map the cognitive landscape of global climate change negotiations and have offered a partial snapshot of the belief systems of negotiators and the effects of these beliefs on the political dynamics of the UNFCCC process. Some of the insights I have presented pertain to a particular moment in the history of climate change diplomacy—the development of the Paris Agreement. Others have more fundamental and lasting implications for both science and politics.

Building on these insights, further work is needed. For example, it would be valuable to identify both additional belief systems that have not been captured by this project and novel belief systems that will doubtless develop over time. Further theoretical work is also needed to understand the fundamental rules that determine the structure and functioning of these belief systems, in particular, the centrality of actor types and identity concepts and their relationship to the nature of the climate governance problem.

Even more important questions for future research concern the nature of cognitive change over time and the processes that can create different kinds of cognitive change in the context of global climate change politics. What are the key variables, patterns, and sequences that lead individuals, communities, or states to change their views? Exploring the relationship between the dynamics of cognitive change, political behavior, and multilateral cooperation might be one of the most valuable areas of research at the intersection of science and global public policy making.

I have identified a number of issues that require attention from science policy and communication experts, futurists, knowledge brokers, and institutional designers. How can we identify and fill knowledge gaps at the science-policy interface, concerning climate tipping points, for example? What kinds of processes could support the scientifically informed imagination of the distant future? These questions concern not only science communicators, but also climate scientists, who make important choices about the phenomena they investigate and the concepts they present in their findings.

I have also pointed to the need to advance current theories of agency in the field of international relations. Building on insights from psychology, sociology, and cognitive science, a political definition of agency should be able to address the person-group problem and be sensitive to the importance of time. Further, the presence or absence of emotions such as hope can play an important role in strengthening or weakening beliefs about agency. Based on my study data, the link between hope and agency in the climate change negotiations created significant biases not just in individual minds, but also in collective processes of selecting global governance goals. Countering such biases requires acknowledging the role of emotions in political behavior and greater methodological creativity to gain some empirical traction on these issues.

Future research could also explore the mutual relationship between the cognitive patterns of climate change negotiators operating at the global scale and actors in domestic politics. Which domestic debates contribute to the belief systems that exist at the global scale? Under what circumstances does a global discourse affect domestic political beliefs? Some of the cognitive patterns identified in *Mindmade Politics*, such as place identity, might also be at work in the minds of domestic policy makers or members of the public.

Finally, and perhaps most important for the field of international relations, I believe there is value in further exploring a cognitive approach as a bridge between realism, institutionalism, and constructivism. Focusing on cognitive processes can create insights about the types of reasoning, motivations, values, and emotions that drive political processes, about their boundary conditions, and about the role of person-group interactions in collective meaning-making processes. Interaction rules between structural and ideational theories are likely to be issue specific, opening up a whole new field of research with multiple case applications outside of climate change.

The Science and Politics of Tears

It was late in the afternoon and the negotiation tents at La Bourget in northern Paris were clearing out. It was the second week of the Paris Conference of the Parties (COP 21), and most conversations were taking place behind closed doors, leaving observers like myself with little to do. I sat down at one of the computers available to all attendees to check my emails. Out of the corner of my eye, I recognized a familiar face looking at the screen to my left. By chance, I had picked a seat next to Yeb Saño. Having seen the crowds that followed him around after Doha and the massive media attention he drew with his Fast for the Climate campaign, it was strange to see him here, all by himself. This man in his loose clothes and worn shoes looked nothing like the diplomat I had known. I asked him how he was doing, and what had happened since he had made a commitment to fast for action on climate change—to fast for all of us. With a tired but

contented smile, he told me about his surprise that multiple faith movements had made him their figurehead, and that he had just walked all the
way from Rome across the Alps to Paris on a pilgrimage to raise awareness
for the climate change movement. Hence the worn shoes.

When Yeb Saño stepped down as the Philippines' climate change com-
missioner and left diplomacy in 2015, diplomacy lost an important voice,
one that was not afraid to say what many others feel. The world moves on,
other diplomats have replaced Saño and his colleagues and there will be
still others to replace them. But perhaps we should pay heed why Yeb Saño
left diplomacy. It was not because he no longer cared or because he was
bad at his job. He left because he stopped believing that diplomacy could
get the job done and because he now placed his trust, not in the UNFCCC
process, but in the ability of social movements to change the world. His
emotions had as much to do with his departure as his thoughts: frustration,
anger, despair, and hope. Negotiators and scholars need to acknowledge
and address not just the thoughts, but also the emotions that will make
our future.

Appendices

Appendix A1   National CO_2 Emissions from Energy Production in 2010 and 2012

Table A.1

Cutoffs between high, medium, and low CO_2 emissions

Total CO_2 emissions from consumption of energy (millions of metric tons)

Country	2010	Rank	Group	Group rank	Totals	Country	2012	Rank	Group	Group rank	Totals
China	8,320.96	1	High	1		China	8,106.43	1	High	1	
United States	5,610.11	2	High	2		United States	5,270.42	2	High	2	
India	1,695.62	3	High	3		India	1,830.94	3	High	3	
Russia	1,633.80	4	High	4		Russia	1,781.72	4	High	4	
Japan	1,164.47	5	High	5		Japan	1,259.06	5	High	5	
Germany	793.66	6	High	6		Germany	788.32	6	High	6	
Korea, South	578.97	7	High	7		Korea, South	657.09	7	High	7	
Iran	560.33	8	High	8		Iran	603.59	8	High	8	
Canada	548.75	9	High	9		Saudi Arabia	582.67	9	High	9	
United Kingdom	532.44	10	High	10		Canada	550.83	10	High	10	
Saudi Arabia	478.41	11	High	11		Brazil	500.23	11	High	11	
South Africa	465.10	12	High	12		United Kingdom	498.88	12	High	12	
Brazil	453.87	13	High	13		South Africa	473.16	13	High	13	
Mexico	445.28	14	High	14		Indonesia	456.21	14	High	14	
Italy	416.37	15	High	15		Mexico	453.83	15	High	15	
Australia	405.34	16	High	16		Australia	420.63	16	High	16	
France	395.20	17	High	17		Italy	385.81	17	High	17	

Table A.1 (continued)

Total CO_2 emissions from consumption of energy (millions of metric tons)

Country	2010	Rank	Group	Group rank	Totals	Country	2012	Rank	Group	Group rank	Totals
Indonesia	389.43	18	High	18		France	364.54	18	High	18	
Spain	316.43	19	High	19		Spain	312.44	19	High	19	
Taiwan	305.38	20	High	20		Taiwan	307.15	20	High	20	
Poland	303.70	21	High	21		Turkey	296.93	21	High	21	
Thailand	278.49	22	High	22		Thailand	290.72	22	High	22	
Ukraine	275.51	23	High	23		Ukraine	290.38	23	High	23	
Turkey	263.54	24	High	24		Poland	289.45	24	High	24	
Netherlands	263.44	25	High	25	25	Netherlands	239.60	25	High	25	
United Arab Emirates	199.37	26	Medium	1		United Arab Emirates	234.06	26	High	26	
Egypt	196.55	27	Medium	2		Kazakhstan	224.22	27	High	27	
Kazakhstan	184.47	28	Medium	3		Singapore	207.96	28	High	28	
Malaysia	181.93	29	Medium	4		Egypt	206.29	29	High	29	29
Singapore	172.19	30	Medium	5		Malaysia	198.79	30	Medium	1	
Argentina	169.83	31	Medium	6		Argentina	196.00	31	Medium	2	
Venezuela	158.44	32	Medium	7		Venezuela	184.79	32	Medium	3	
Pakistan	151.65	33	Medium	8		Pakistan	146.89	33	Medium	4	
Belgium	127.19	34	Medium	9		Belgium	139.14	34	Medium	5	
Iraq	118.31	35	Medium	10		Algeria	133.92	35	Medium	6	
Uzbekistan	114.27	36	Medium	11		Vietnam	131.73	36	Medium	7	

Table A.1 (continued)

Total CO_2 emissions from consumption of energy (millions of metric tons)

Country	2010	Rank	Group	Group rank	Totals	Country	2012	Rank	Group	Group rank	Totals
Turkmenistan	62.05	55	Medium	30		Oman	62.85	55	Medium	26	
Libya	60.60	56	Medium	31		Libya	54.60	56	Medium	27	
Bangladesh	56.74	57	Medium	32		Peru	53.58	57	Medium	28	
Oman	55.20	58	Medium	33		Trinidad and Tobago	51.27	58	Medium	29	
Finland	54.40	59	Medium	34		Portugal	51.20	59	Medium	30	
Portugal	51.43	60	Medium	35		Sweden	51.08	60	Medium	31	
Hungary	50.39	61	Medium	36		Syria	50.92	61	Medium	32	
Trinidad and Tobago	49.93	62	Medium	37		Bulgaria	48.85	62	Medium	33	
Serbia	49.92	63	Medium	38		Hungary	47.90	63	Medium	34	
Denmark	45.96	64	Medium	39		Finland	46.81	64	Medium	35	
Switzerland	45.55	65	Medium	40		Switzerland	42.97	65	Medium	36	
Bulgaria	42.17	66	Medium	41		Serbia	41.38	66	Medium	37	
Peru	41.88	67	Medium	42		Norway	41.06	67	Medium	38	
Norway	41.80	68	Medium	43		Denmark	40.51	68	Medium	39	
Ireland	40.48	69	Medium	44		Morocco	39.35	69	Medium	40	
New Zealand	39.58	70	Medium	45		New Zealand	37.89	70	Medium	41	
Morocco	35.66	71	Medium	46		Ecuador	37.23	71	Medium	42	
Azerbaijan	35.12	72	Medium	47		Ireland	35.49	72	Medium	43	

Table A.1 (continued)

Total CO_2 emissions from consumption of energy (millions of metric tons)

Country	2010	Rank	Group	Group rank	Totals	Country	2012	Rank	Group	Group rank	Totals
Slovakia	34.54	73	Medium	48		Azerbaijan	35.14	73	Medium	44	
Cuba	34.46	74	Medium	49		Bahrain	32.20	74	Medium	45	
Puerto Rico	30.86	75	Medium	50		Slovakia	32.08	75	Medium	46	
Bahrain	30.69	76	Medium	51		Angola	31.61	76	Medium	47	
Yemen	26.50	77	Medium	52		Puerto Rico	26.81	77	Medium	48	
Ecuador	24.43	78	Medium	53		Bosnia and Herzegovina	26.00	78	Medium	49	
Angola	24.20	79	Medium	54		Cuba	25.99	79	Medium	50	
Croatia	23.43	80	Medium	55		Yemen	21.28	80	Medium	51	
Estonia	20.56	81	Medium	56		Dominican Republic	20.80	81	Medium	52	
Bosnia and Herzegovina	20.14	82	Medium	57		Tunisia	20.27	82	Medium	53	
Dominican Republic	19.60	83	Medium	58		Croatia	20.18	83	Medium	54	
Jordan	19.07	84	Medium	59		Montenegro	19.72	84	Medium	55	
Tunisia	18.72	85	Medium	60		Bolivia	17.28	85	Medium	56	
Slovenia	17.42	86	Medium	61		Jordan	16.86	86	Medium	57	
Lithuania	15.98	87	Medium	62		Lithuania	16.69	87	Medium	58	
Panama	15.46	88	Medium	63		Lebanon	16.44	88	Medium	59	
Lebanon	15.24	89	Medium	64		Panama	16.23	89	Medium	60	
Sri Lanka	14.09	90	Medium	65		Slovenia	15.87	90	Medium	61	
Sudan and South Sudan	13.79	91	Medium	66		Sri Lanka	15.23	91	Medium	62	

Table A.1 (continued)

Total CO_2 emissions from consumption of energy (millions of metric tons)

Country	2010	Rank	Group	Group rank	Totals	Country	2012	Rank	Group	Group rank	Totals
Vietnam	112.80	37	Medium	12		Iraq	130.74	37	Medium	8	
Algeria	110.90	38	Medium	13		Uzbekistan	123.17	38	Medium	9	
Greece	92.99	39	Medium	14		Kuwait	105.68	39	Medium	10	
Czech Republic	90.83	40	Medium	15		Qatar	99.17	40	Medium	11	
Philippines	85.63	41	Medium	16		Czech Republic	91.15	41	Medium	12	
Hong Kong	83.78	42	Medium	17		Hong Kong	88.63	42	Medium	13	
Kuwait	81.33	43	Medium	18		Greece	87.56	43	Medium	14	
Nigeria	80.51	44	Medium	19		Nigeria	86.40	44	Medium	15	
Romania	78.43	45	Medium	20		Romania	86.06	45	Medium	16	
Colombia	72.31	46	Medium	21		Philippines	83.95	46	Medium	17	
Israel	70.32	47	Medium	22		Chile	81.51	47	Medium	18	
Austria	69.46	48	Medium	23		Israel	80.36	48	Medium	19	
Chile	68.76	49	Medium	24		Colombia	74.90	49	Medium	20	
Belarus	68.24	50	Medium	25		Belarus	67.13	50	Medium	21	
Qatar	64.68	51	Medium	26		Korea, North	67.00	51	Medium	22	
Korea, North	63.69	52	Medium	27		Austria	66.68	52	Medium	23	
Syria	63.10	53	Medium	28		Turkmenistan	64.98	53	Medium	24	
Sweden	62.74	54	Medium	29		Bangladesh	63.50	54	Medium	25	

Table A.1 (continued)

Total CO_2 emissions from consumption of energy (millions of metric tons)

Country	2010	Rank	Group	Group rank	Totals	Country	2012	Rank	Group	Group rank	Totals
Bolivia	13.29	92	Medium	67		Sudan and South Sudan	13.94	92	Medium	63	
Guatemala	12.97	93	Medium	68		Kenya	13.45	93	Medium	64	
Burma (Myanmar)	12.80	94	Medium	69		Burma (Myanmar)	13.34	94	Medium	65	
Kenya	12.25	95	Medium	70		Guatemala	13.07	95	Medium	66	
Virgin Islands, U.S.	11.95	96	Medium	71		Jamaica	12.75	96	Medium	67	
Armenia	11.56	97	Medium	72		Virgin Islands, U.S.	12.41	97	Medium	68	
Luxembourg	10.80	98	Medium	73		Armenia	12.12	98	Medium	69	
Ghana	10.58	99	Medium	74	74	Netherlands Antilles	11.84	99	Medium	70	
Mongolia	9.44	100	Low	1		Luxembourg	11.69	100	Medium	71	
Cyprus	9.26	101	Low	2		Mongolia	11.36	101	Medium	72	
Jamaica	9.22	102	Low	3		Honduras	10.33	102	Medium	73	
Latvia	9.07	103	Low	4		Zimbabwe	10.12	103	Medium	74	74
Netherlands Antilles	8.82	104	Low	5		Moldova	9.41	104	Low	1	
Zimbabwe	8.49	105	Low	6		Tanzania	9.30	105	Low	2	
Honduras	8.29	106	Low	7		Kyrgyzstan	9.28	106	Low	3	
Brunei	8.27	107	Low	8		Ghana	9.10	107	Low	4	
Macedonia	8.23	108	Low	9		Cyprus	8.80	108	Low	5	
Tanzania	7.57	109	Low	10		Brunei	8.68	109	Low	6	
Moldova	7.38	110	Low	11		Afghanistan	8.55	110	Low	7	

Table A.1 (continued)

Total CO$_2$ emissions from consumption of energy (millions of metric tons)

Country	2010	Rank	Group	Group rank	Totals	Country	2012	Rank	Group	Group rank	Totals
Cameroon	7.36	111	Low	12		Ethiopia	8.21	111	Low	8	
Uruguay	7.27	112	Low	13		Macedonia	8.08	112	Low	9	
Ethiopia	6.74	113	Low	14		Latvia	7.90	113	Low	10	
Senegal	6.68	114	Low	15		Uruguay	7.59	114	Low	11	
Tajikistan	6.68	115	Low	16		Kosovo	7.58	115	Low	12	
Congo (Brazzaville)	6.52	116	Low	17		Costa Rica	7.29	116	Low	13	
El Salvador	6.48	117	Low	18		Senegal	7.14	117	Low	14	
Costa Rica	6.41	118	Low	19		Congo (Brazzaville)	6.69	118	Low	15	
Côte d'Ivoire (Ivory Coast)	5.94	119	Low	20		Malta	6.56	119	Low	16	
Bahamas	5.57	120	Low	21		Côte d'Ivoire (Ivory Coast)	6.40	120	Low	17	
Papua New Guinea	5.31	121	Low	22		El Salvador	6.37	121	Low	18	
Georgia	5.30	122	Low	23		Georgia	6.26	122	Low	19	
Equatorial Guinea	5.00	123	Low	24		Cameroon	6.22	123	Low	20	
Albania	4.89	124	Low	25		Cambodia	6.05	124	Low	21	
Nicaragua	4.82	125	Low	26		Estonia	5.69	125	Low	22	
Gabon	4.59	126	Low	27		Equatorial Guinea	5.61	126	Low	23	
Mauritius	4.55	127	Low	28		Gabon	5.44	127	Low	24	

Table A.1 (continued)

Total CO_2 emissions from consumption of energy (millions of metric tons)

Country	2010	Rank	Group	Group rank	Totals	Country	2012	Rank	Group	Group rank	Totals
Paraguay	4.39	128	Low	29		Mauritius	5.32	128	Low	25	
Kyrgyzstan	4.13	129	Low	30		Nicaragua	5.29	129	Low	26	
Gibraltar	3.91	130	Low	31		Mozambique	4.79	130	Low	27	
Botswana	3.84	131	Low	32		Benin	4.58	131	Low	28	
Namibia	3.81	132	Low	33		Albania	3.96	132	Low	29	
Benin	3.65	133	Low	34		Gibraltar	3.95	133	Low	30	
Cambodia	3.59	134	Low	35		Botswana	3.92	134	Low	31	
Madagascar	3.38	135	Low	36		Paraguay	3.87	135	Low	32	
Palestine	3.38	136	Low	37		Bahamas	3.84	136	Low	33	
Nepal	3.36	137	Low	38		Namibia	3.72	137	Low	34	
Iceland	3.36	138	Low	39		Nepal	3.64	138	Low	35	
Togo	3.17	139	Low	40		Iceland	3.50	139	Low	36	
Malta	3.11	140	Low	41		Papua New Guinea	3.38	140	Low	37	
New Caledonia	3.03	141	Low	42		New Caledonia	3.07	141	Low	38	
Reunion	2.96	142	Low	43		Zambia	3.05	142	Low	39	
Mauritania	2.89	143	Low	44		Palestinian Territories	3.01	143	Low	40	
Congo (Kinshasa)	2.80	144	Low	45		Tajikistan	2.97	144	Low	41	
Martinique	2.77	145	Low	46		Madagascar	2.89	145	Low	42	
Mozambique	2.73	146	Low	47		Martinique	2.84	146	Low	43	

Table A.1 (continued)

Total CO_2 emissions from consumption of energy (millions of metric tons)

Country	2010	Rank	Group	Group rank	Totals	Country	2012	Rank	Group	Group rank	Totals
Macau	2.54	147	Low	48		Reunion	2.77	147	Low	44	
Fiji	2.50	148	Low	49		Uganda	2.55	148	Low	45	
Zambia	2.42	149	Low	50		Congo (Kinshasa)	2.48	149	Low	46	
Djibouti	2.35	150	Low	51		Mauritania	2.41	150	Low	47	
Suriname	2.34	151	Low	52		Guadeloupe	2.33	151	Low	48	
Guadeloupe	2.10	152	Low	53		Suriname	2.27	152	Low	49	
Uganda	2.01	153	Low	54		Haiti	2.09	153	Low	50	
Montenegro	1.94	154	Low	55		Malawi	1.91	154	Low	51	
Niger	1.80	155	Low	56		Djibouti	1.80	155	Low	52	
Barbados	1.57	156	Low	57		Guam	1.77	156	Low	53	
Guyana	1.52	157	Low	58		Macau	1.69	157	Low	54	
Haiti	1.46	158	Low	59		Guyana	1.66	158	Low	55	
Guam	1.45	159	Low	60		Togo	1.63	159	Low	56	
Burkina Faso	1.44	160	Low	61		Laos	1.62	160	Low	57	
Guinea	1.39	161	Low	62		Fiji	1.54	161	Low	58	
Malawi	1.36	162	Low	63		Niger	1.41	162	Low	59	
Sierra Leone	1.33	163	Low	64		Burkina Faso	1.41	163	Low	60	
Wake Island	1.29	164	Low	65		Guinea	1.39	164	Low	61	
Seychelles	1.25	165	Low	66		Barbados	1.31	165	Low	62	

Table A.1 (continued)

Total CO_2 emissions from consumption of energy (millions of metric tons)

Country	2010	Rank	Group	Group rank	Totals	Country	2012	Rank	Group	Group rank	Totals
Aruba	1.24	166	Low	67		Sierra Leone	1.31	166	Low	63	
French Polynesia	1.22	167	Low	68		Seychelles	1.30	167	Low	64	
Laos	1.19	168	Low	69		Wake Island	1.29	168	Low	65	
French Guiana	1.12	169	Low	70		Maldives	1.12	169	Low	66	
Swaziland	1.11	170	Low	71		French Polynesia	1.07	170	Low	67	
Belize	0.98	171	Low	72		French Guiana	1.04	171	Low	68	
Maldives	0.92	172	Low	73		Swaziland	0.94	172	Low	69	
Somalia	0.90	173	Low	74		Aruba	0.88	173	Low	70	
Mali	0.89	174	Low	75		Somalia	0.86	174	Low	71	
Rwanda	0.87	175	Low	76		Mali	0.77	175	Low	72	
Bermuda	0.81	176	Low	77		Rwanda	0.77	176	Low	73	
Eritrea	0.80	177	Low	78		Faroe Islands	0.75	177	Low	74	
Afghanistan	0.79	178	Low	79		Eritrea	0.74	178	Low	75	
Faroe Islands	0.74	179	Low	80		Belize	0.68	179	Low	76	
Liberia	0.74	180	Low	81		Bermuda	0.61	180	Low	77	
Antigua and Barbuda	0.72	181	Low	82		American Samoa	0.61	181	Low	78	
American Samoa	0.69	182	Low	83		Greenland	0.60	182	Low	79	
Greenland	0.56	183	Low	84		Antigua and Barbuda	0.59	183	Low	80	
Burundi	0.53	184	Low	85		Liberia	0.54	184	Low	81	

Table A.1 (continued)

Total CO_2 emissions from consumption of energy (millions of metric tons)

Country	2010	Rank	Group	Group rank	Totals	Country	2012	Rank	Group	Group rank	Totals
Guinea-Bissau	0.46	185	Low	86		Timor-Leste (East Timor)	0.50	185	Low	82	
Saint Lucia	0.43	186	Low	87		Cayman Islands	0.47	186	Low	83	
Grenada	0.43	187	Low	88		Gambia, The	0.47	187	Low	84	
Timor-Leste (East Timor)	0.40	188	Low	89		Guinea-Bissau	0.46	188	Low	85	
Solomon Islands	0.36	189	Low	90		Central African Republic	0.44	189	Low	86	
Saint Vincent/Grenadines	0.33	190	Low	91		Grenada	0.43	190	Low	87	
Western Sahara	0.31	191	Low	92		Saint Lucia	0.42	191	Low	88	
Antarctica	0.31	192	Low	93		Cape Verde	0.39	192	Low	89	
Saint Kitts and Nevis	0.30	193	Low	94		Bhutan	0.32	193	Low	90	
Cook Islands	0.30	194	Low	95		Western Sahara	0.32	194	Low	91	
U.S. Pacific Islands	0.29	195	Low	96		Burundi	0.32	195	Low	92	
Gambia	0.29	196	Low	97		U.S. Pacific Islands	0.29	196	Low	93	
Chad	0.29	197	Low	98		Lesotho	0.27	197	Low	94	
Lesotho	0.28	198	Low	99		Saint Vincent/Grenadines	0.27	198	Low	95	
Bhutan	0.28	199	Low	100		Solomon Islands	0.27	199	Low	96	
Cayman Islands	0.27	200	Low	101		Chad	0.26	200	Low	97	
Cape Verde	0.27	201	Low	102		Saint Kitts and Nevis	0.25	201	Low	98	
Central African Republic	0.23	202	Low	103		Tonga	0.19	202	Low	99	

Table A.1 (continued)

Total CO_2 emissions from consumption of energy (millions of metric tons)

Country	2010	Rank	Group	Group rank	Totals	Country	2012	Rank	Group	Group rank	Totals
Nauru	0.22	203	Low	104		Nauru	0.17	203	Low	100	
Tonga	0.16	204	Low	105		Vanuatu	0.17	204	Low	101	
Saint Pierre and Miquelon	0.15	205	Low	106		Samoa	0.16	205	Low	102	
São Tomé and Principe	0.15	206	Low	107		Virgin Islands, British	0.16	206	Low	103	
Comoros	0.15	207	Low	108		Turks and Caicos Islands	0.16	207	Low	104	
Virgin Islands, British	0.15	208	Low	109		Comoros	0.16	208	Low	105	
Samoa	0.15	209	Low	110		Saint Pierre and Miquelon	0.15	209	Low	106	
Montserrat	0.15	210	Low	111		Cook Islands	0.15	210	Low	107	
Vanuatu	0.15	211	Low	112		São Tomé and Principe	0.14	211	Low	108	
Dominica	0.14	212	Low	113		Dominica	0.13	212	Low	109	
Falkland Islands	0.05	213	Low	114		Antarctica	0.10	213	Low	110	
Kiribati	0.04	214	Low	115		Montserrat	0.09	214	Low	111	
Turks and Caicos Islands	0.04	215	Low	116		Kiribati	0.06	215	Low	112	
Saint Helena	0.02	216	Low	117		Falkland Islands	0.05	216	Low	113	
Niue	0.01	217	Low	118	118	Saint Helena	0.01	217	Low	114	
						Niue	0.00	218	Low	115	115
World	31,780.36					World	32,310.29				

Source: U.S. Energy Information Administration, http://www.eia.gov.

Appendix A2   Study Participants' Membership in Current UNFCCC Negotiation Groups

Totals refer to the total number of study participants in a given negotiation group (e.g., only five countries of the Umbrella Group are represented, but eight individuals from these five countries participated in the study.)

Table A.2
Participants' membership in current UNFCCC negotiation groups

State	Negotiation group										Parti-cipant group
	Umbrella Group	European Union	G77 and China	AOSIS, SIDS	BASIC	LMDC	LDCs	ALBA	EIG	CRN	
1 Argentina			✓							✓	5-ML
2 Australia	✓										4-HL
3 Bangladesh			✓				✓			✓	2-MH
4 Barbados			✓	✓							6-LL
5 Bolivia						✓		✓			2-MH
6 Botswana			✓								3-LH
7 Brazil			✓		✓						1-HH
8 Canada (2)	✓										4-HL
9 Cape Verde			✓	✓							6-LL
10 Colombia			✓								2-MH
11 Denmark		✓									5-ML
12 Dominica			✓	✓						✓	3-LH
13 Finland (2)		✓									5-ML
14 Germany		✓	✓								4-HL
15 Grenada			✓	✓							3-LH
16 Guatemala			✓							✓	2-MH

Table A.2 (continued)

State	Negotiation group										Participant group
	Umbrella Group	European Union	G77 and China	AOSIS, SIDS	BASIC	LMDC	LDCs	ALBA	EIG	CRN	
17 Iceland	✓										6-LL
18 Indonesia			✓			✓				✓	1-HH
19 Japan	✓										4-HL
20 Mozambique			✓				✓				3-LH
21 Namibia			✓								3-LH
22 Pakistan			✓			✓				✓	2-MH
23 Philippines			✓			✓					2-MH
24 Samoa			✓	✓			✓			✓	3-LH
25 Singapore				✓							5-ML
26 South Africa			✓		✓						1-HH
27 South Korea (2)									✓		4-HL
28 Sweden (2)		✓									5-ML
29 Uganda			✓				✓				3-LH
30 United States (3)	✓										4-HL
Totals*	8	5	18	6	2	5	4	1	2	8	

*Totals refer to the total number of study participants in given negotiation group.

Appendix A3   Cognitive-Affective Mapping: Protocol and Limitations

Cognitive-affective mapping focuses on one specific aspect of human cognition: the relevance of emotions attached to key values and beliefs that individuals hold with respect to a certain issue. To that end, CAMs are highly simplified representations of the relationships of important concepts and associated emotions. They are not supposed or intended to resemble a neurological structure or process or any other biological reality. Nor do they provide a full picture of the mind, or even of the part of the mind that is relevant for the topic under discussion. Other tools and models are more appropriate for more comprehensive representations (and simulations) of the multiple interacting facets of cognition. Given the very specific purpose of integrating emotion into a conceptual structure, the CAM tool has a number of important limitations that constrain what it can represent and how. The process of generating a CAM based on interview transcripts often runs into these limitations without being able to resolve them fully. Below I summarize the most relevant limitations and indicate the standard protocol I used when faced with these problems in the course of this project.

Different Relationships Between Two Nodes

Although CAM links represent primarily an emotional relationship between two concepts—likeness or opposition—they also indicate cognitive-semantic relationships. In fact, the emotional valences are rooted in and depend on a cognitive-semantic link between two concepts. But CAM links can also represent very different types of cognitive-semantic relationships, such as constitution, causation, correlation, association, (inter-) dependence, and any form of influence (e.g., to support, increase, strengthen, weaken, limit, decrease, accelerate, damage). Third-party viewers of a CAM might not have enough information to identify the meaning of any particular link. Although much can be inferred from the conceptual context, the viewers may have to insert their own assumptions regarding the meaning of the link, rather than simply "seeing" conceptual structures. This problem can be addressed by expanding the CAM to add concepts that differentiate and clarify the relationship.

Using the representational tools offered by cognitive-affective mapping, at least three types of cognitive-semantic links are possible with different emotional implications. The first and most important one is emotional association. If two concepts have the same emotional valence, in other words, if a person (or group) likes one, the person (or group) also likes the other, the link is represented as a solid line. Two emotionally different, or opposed, concepts are represented with a dashed line (i.e., the person dislikes the link). This is the default way of representing links.

Example: Somebody states that climate change is a threat to humanity.
[Climate Change]———[Threat]--------(Humanity)

Since "climate change" and "threat" are both negative concepts, their link is emotionally similar, but the relationship between "threat" (negative) and "humanity" (positive) is one of emotional opposition.

The remaining two types of cognitive-semantic links between two concepts are of a causal or influential nature. In a "positive causality" link, one concept causes, constitutes, or positively influences the other, whereas in a "negative causality" link, one concept negatively affects the other. Adding the emotional valences of the nodes creates four different options: (1) two nodes with the same emotional valence have a positive causal relationship (solid line); (2) two positive nodes have a negative causal relationship (dotted line); (3) a positive and a negative node have a positive causal relationship (solid line); and (4) a positive and a negative node have a negative causal relationship (dotted line). Option 1 cannot be visually distinguished from emotional coherence and is unproblematic because it matches the emotional information. Option 4 is visually the same as emotional incoherence and is also unproblematic because cognitive and emotional information are aligned.

Example for option 2:

I would like to have a successful career, but also lots of time with my family.
(Successful Career)·········(Family Time)

Example for option 3:

Natural disasters will increase public attention to climate change, which is necessary to change current policies.
[Natural Disasters]———(Public Attention)

In this case, the causal effect of natural disasters on public attention is desirable and outweighs the emotional incoherence. Alternatively, one can depict the concept "natural disasters" as ambivalent.

Although the CAMs in this study generally prioritized emotional associations between nodes over causal ones, in circumstances when the available interview data prioritized causal connections, I chose to represent causality when it was necessary to create a comprehensible CAM.

Directionality

CAM networks are undirected (in other words, they imply a mutual influence), which is one of the major differences of cognitive-affective mapping from other conceptual mapping approaches. Combined with the problem described above (multiple possible relationships), this can create significant difficulties when creating and reading CAMs. Depending on the type of relationship between nodes, there could be unidirectionality (e.g., causation), mutual constitution (i.e., the same kind of influence operating in both directions), or different types of links between the same two nodes (e.g., a constitutional relationship in one direction; a negative influence in the other).

The problem of unidirectionality (e.g., Mary loves John) cannot be resolved given the symmetric links of CAMs. The concept "loves" would be connected to both "John" and "Mary," leaving viewers of the CAM unclear whether Mary loves John or John loves Mary. If the conceptual context does not provide the needed clarification, the researcher needs to provide the necessary interpretation.

Based on these concerns, it is useful to distinguish between the character of the local relationship between two nodes, on the one hand, and the character of a larger set of CAM elements as coherent or incoherent, on the other. Coherence is an appropriate global descriptor for a macro-level state that emerges from the connections between multiple nodes (a cluster of nodes or the entire CAM), but not necessarily the most appropriate term for the relationship between two nodes.

Multiple Representation Options for the Same Thought Structure

Theoretically, there are multiple ways of representing the same cognitive structure, which introduces an element of researcher-driven randomness.

Example:

I like peace and dislike war; peace and war are incoherent with each other.
(Peace)——[War]
I don't like the absence of peace and I don't like war; both concepts are coherent with each other.
[No Peace]——[War]

Most conceptual relationships can be represented in at least two different ways, using inversion or opposites. Since CAMs do not claim to represent a real mental structure or rules of language processing, this feature is not problematic as long the meaning and content of the individual's thoughts are successfully depicted.

The standard protocol in this study was to closely adhere to the language used by study participants, respecting their choice of concepts.

The Use of Opposites

The last point raises the question to what extent it is useful to include opposite concepts in a CAM. If one assumes that people are logically consistent, one could argue that opposites are superfluous information. This issue is particularly relevant for representing decision making when the choice is between "Do A" and "Don't do A." The concepts related to each of these decision options might differ, although they are merely opposites, for example, "Doing A makes me happy and allows me to meet three old friends, whereas not doing A does not make me unhappy and allows me to hear an interesting talk." Therefore it can be useful to include them both when analyzing a decision-making process.

This study did not map decision-making processes and did not face any problems with opposite concepts.

Representing Temporal, Conditional, and Normative Beliefs

The key strength of CAMs is their ability to represent the relationships of values, which are related to each other in a time-independent, factual manner—the world as it is (the present state of affairs) from the perspective of the person or group whose views are subject of the CAM. It is also strongly suited to displaying a person's worldview or ideology—the system of connected concepts and ideas that tend to be stable over long periods of time and that are independent of the specifics of a situation or policy problem. This descriptive focus on present relationships of values cannot readily represent the passing of time (e.g., concepts associated with the past or future and their temporal connection), conditional statements (if-then), indirect beliefs (beliefs attributed to others), or normative beliefs (e.g., "The rich ought to take responsibility for the poor"). Further, a single CAM cannot display changes in a person's or group's views over time.

Creating separate CAMs, for example, the "is" world versus the "ought" world, the world in 1997 versus the world in 2012, "my" world versus "my mother's" world, could solve some of these issues.

In this study, I have compressed all these issues into a single CAM for each study participant and have indicated temporal differences with separate nodes for years, or a node for "past" and "today." Other problems, especially normative statements require verbal interpretation.

Negation

Because CAMs cannot represent negation or negative definitions (i.e., what something is not), they would indicate the belief that concept X (e.g., climate change) is not concept Y (e.g., a threat to humanity) simply by the *absence* of a link between these two concepts.

Lack of Data

Although CAMs seek to depict the emotional intensity of nodes and the strength of links, interviews do not provide reliable data for these depictions.

The default procedure for this study was to assign a weight of +1 or –1 for each concept or link and to add emotional strength only when additional indicators in the interview transcript were available (e.g., adjectives like "horrible," "terrifying," "scary").

Appendix A4   Interview Protocol

1. What was your role during the UNFCCC negotiations at COP 17 in Durban/COP 3 in Kyoto?

Part 1: The Nature of the Climate Change Problem

2. How would you summarize the most important and well-established scientific facts about climate change?

3. Which future climate change impacts are you most concerned about? When do you expect "serious" climate impacts to occur? What would that look like?

4. Do you think we are already seeing climate change effects today?

5. Do you think about climate change mainly as a threat to your country's citizens/organization's members or others?

6. Do you think about climate change mainly as a threat to people today or future generations?

7. Thinking about climate change as a political rather than scientific issue, how would you describe the nature of the problem?

8. Do you believe it is possible to address climate change effectively? And what does it mean to "solve the climate problem"? Who has the ability to address the problem? How?

9. What is the goal of the negotiations? Is the 2°C target a good goal?

Part 2: Negotiation Positions and Durban Negotiations

10. When preparing for the negotiations, how do you define your organization's (national) interest? Did your negotiation preparation contain some form of a cost-benefit analysis?

11. What are the main obstacles to finding an effective international agreement on climate change?

12. Do you think that Copenhagen/Durban was a success or failure?

13. Was Copenhagen/Durban an emotional experience for you? What are the most important emotional situations you remember? How did you feel?

Part 3: Special Characteristics of Climate Change

14. Are you concerned about climate tipping points? Does the possibility of tipping points influence how you think about the negotiations?

Part 4: Imagining the Future

15. Please imagine that all efforts to create an effective multilateral agreement on climate change fail. What would the world look like in a worst-case climate change scenario in the year 2080?

Appendix A5   Cognitive-Affective Maps: Coding Scheme

Table A.3
Coding scheme of cognitive-affective maps

Approach	Theme	Current categories	Subcategories
Part 1: International relations theory	Rational-choice institutionalism	Rational choice	Costs
			Benefits
			Power balancing
			National interest
			Institutional effects
			Incentive Structure
	Social constructivism	Identity	Group membership
			In- vs. out-group/ in-group preference
			Group purpose
			Group norms (nonjustice)
			Place-identity
		Norms and justice	Group norms (justice)
			Equity
			North–South divide
			Fairness
			Other
	Domestic politics	Two-level game	Domestic actors
			Domestic processes
			Diplomatic mandates
Part 2: Special characteristics	Complexity	Systemic thinking	Nonlinearity and feedbacks
			Multiple scales
			Cascading effects
			Other

Table A.3 (continued)

Approach	Theme	Current categories	Subcategories
		Mechanistic thinking	Incrementalism
			Linearity
		Pervasiveness	Societal change
			Everybody's responsibility
	Uncertainty	Scientific	
		General	
		Measuring success	
	Myopia	Time scales	Peaking
			ADP Agreement
			Delay/Inertia
			Future generations
			Other
		Institutionalized short-termism	Democratic elections
			Other
	Imperceptibility	Abstract knowledge	
		Personal experience	
		Third-party experience	
	Climate tipping points	Irreversibility	
		Catastrophe	
	Multiple stresses		
	Lack of intentionality	Blame	Historical responsibility
Other	Agency	Actor types	States
			International community
			Individual
			Market
			Other

Table A.3 (continued)

Approach	Theme	Current categories	Subcategories
		Paths of influence	Top-down (regulation, international agreement)
			Bottom-up (voluntary standards)
			Other
		Model of change	Incremental vs. radical/transformational
			Driver (policy, technology, market, behavior, values)
			Sequence of changes
			Other
	Hope	Sources	
		Strength	
		Loss of hope	
	Emotions	Positive	(e.g., pride, excitement, hope)
		Negative	(e.g., frustration, shame, anger)
		Other	
	Goal	2°C	
		Climate stability	
		Human well-being	
		Security/Safety	
		Health	
		Food security	
		Other	

Appendix A6   Q Method: List of Statements (Q Set)

Causes of Climate Change

1. Human-released greenhouse gases are causing significant climate change.
2. I don't trust what scientists say about climate change.
3. Anti–climate change policies threaten progress and modernity.

Consequences of Climate Change

4. I do not believe that we will see significant effects of climate change in my lifetime.
5. The social consequences of climate change are most worrying: Whole countries and cultures will disappear; many people will suffer from food and water scarcity, or will be forced to migrate.
6. Climate change is mainly an issue for the developing countries.
7. Climate change was caused by rich, industrialized countries, but its impacts will hurt poor countries most—this asymmetry is unfair.
8. Climate change will result in violence and human deaths.

Actors and Agency

9. An effective climate solution requires the cooperation of all governments around the world.
10. States are the most important players in global climate politics.
11. Even if a new international treaty is signed with ambitious mitigation targets, many states will simply not be able to implement these targets domestically.
12. Only a small number of countries with significant GHG emissions are important for climate negotiations.
13. All our efforts and resources to combat climate change should be concentrated domestically, rather than internationally.
14. The power disparities between countries in climate negotiations make me very upset.
15. The climate problem should be left to the markets.
16. It is best to leave the development of climate solutions to regions, cities, and local communities—they have been much more successful than UNFCCC negotiations.
17. Individual contributions don't make a difference when it comes to climate change.

18. Politicians need much stronger voter support and pressure from political movements to create meaningful climate policies.
19. The vested interests blocking solutions are too powerful to allow for any meaningful action on climate change.
20. Nothing will happen before a climate crisis hits.
21. Neither states nor markets nor civil societies can solve this problem on their own—climate change is a multilevel problem and requires action at all of these different levels.
22. A key element in solving the climate problem is the need for fundamental value change within our societies.
23. Based on our shared humanity, our desire for happiness and security, we can find a solution to the climate problem.
24. God has made us stewards of the Earth, giving us the ability and responsibility to keep the planet healthy.

Policy Goals and Options

25. The prospect of a cleaner, eco-friendly economy is exciting.
26. Limiting average global warming to 2°C will be sufficient to prevent major damage.
27. We lack a clearly defined goal for global climate policy.
28. Since the climate is going to change we should be more concerned with adaptation.
29. Adaptation and mitigation are complementary and equally important policies.
30. Given the political gridlock on mitigation, we might have to resort to geo-engineering to buy more time for mitigation and adaptation.
31. Geo-engineering can solve the global warming problem much more cost-effectively than mitigation.
32. Taxes—whether globally or domestically—and other policies that constrain private property rights are simply not politically acceptable in our system.

Climate Economics and Development

33. Economic growth and jobs must take priority over climate concerns.
34. Ideally, climate policies would reduce GHG emissions while stimulating economic growth.
35. Climate change policy should be based on cost-benefit analysis.
36. The main costs of climate change policies include loss of GDP and jobs.

37. The main costs of future climate change can simply not be calculated: the loss of human life, food insecurity, or species extinctions don't have price tags.
38. Economic growth is the best solution to climate change.
39. Investment in climate policies is a poor use of our resources; it makes more sense to do something about poverty, health care, and education in the developing world.

Identity

40. The prospect of major environmental change, such as the melting of glaciers or species extinction, is very distressing.
41. I am ashamed that my country is not doing more about climate change.
42. The BASIC countries (Brazil, South Africa, India, and China) should show greater leadership in international climate negotiations.
43. I dedicate my career to solving the climate problem.
44. It is disappointing to see how little governments and markets care about the environment and the health of the planet.
45. My government is a very constructive player in international affairs.

Ethics and Justice

46. Elected officials have a political responsibility to protect the interests of their constituency—the present rather than future generations.
47. Problems that might arise decades from now are not important to me.
48. All states have a moral responsibility to contribute to a global climate solution.
49. The current generation (of politicians and voters) has a major ethical responsibility to future generations.
50. Future generations are likely to be richer and better off than we are, and better able to deal with climate change.
51. I fully support climate funding—financial flows from the rich to the poor to help them cope with climate change.
52. Rich countries have caused the problem; consequently, they have the obligation to fix it.

Special Characteristics: Hope, Uncertainty, Myopia, Imperceptibility, Climate Tipping Points

53. Contemplating climate change usually leaves me feeling rather helpless.

54. Climate change makes me fear for my children's future.

55. I believe that we will find a cooperative solution to climate change. Other issues have taken many years of negotiation, too.

56. Climate change is a very depressing issue.

57. Climate change is simply too complex and overwhelming. It is impossible to fully understand, let alone solve, the problem.

58. It is already too late to do anything about climate change.

59. There are moments when I lose all hope that the UNFCCC process can solve this problem.

60. Climate change scares me because I don't know what's going to happen.

61. Climate change is not the only issue we have to deal with and other issues are often more urgent.

62. The focus on winning the next election is the biggest obstacle to finding international agreement.

63. I sometimes wonder how to explain our failure to fix the climate problem to my grandchildren.

64. I have a hard time imagining the consequences of climate change for my community and my country.

65. Sometimes gradual processes such as GHG emissions result in sudden, dramatic changes in the environment. The existence of such climate tipping points makes action even more urgent than previously thought. Avoiding tipping points should become a key climate policy goal.

Appendix A7   Q Method: Instructions and Questions

Instructions

Step 1: Reading and Presorting Statements

In a moment, you will be shown a set of opinion statements (on climate change and international cooperation) on sixty-five "cards." These statements have been drawn from a broad set of sources and reflect various perspectives on the climate issue. The statements show up in a random order, but they fall in roughly eight different categories: (1) climate science, (2) nature of the problem, (3) actors and agency, (4) policy goals and options, (5) economics and development, (6) identity, (7) ethics and justice, and (8) special characteristics. Later, you will be asked to what extent you agree or disagree with these statements on a scale from –5 to +5.

Please read the statements on the cards carefully and split them up into three piles: a pile for statements you tend to disagree with (on the left), a pile for statements you tend to agree with (on the right), and a pile for the rest (in the middle). The middle pile should contain statements that are not very meaningful to you or that you feel conflicted about (you cannot decide whether you agree or disagree with them).

Do not worry about the number of statements in each pile. The distribution is not relevant. For example, you could place 10 statements in the "agree" pile, 12 in the "uncertain" pile, and the rest in the "disagree" pile.

You can either drag the cards into one of the three piles or press 1, 2, 3 on your keyboard. Changes can be made later.

You might find it useful to work with a sheet of paper and note down the values (between –5 and +5) you might later assign to these statements. You could also note your thoughts on statements you find hard to place. At the end of the survey, you will be asked about those.

Please note the following abbreviations and definitions:

GHG: Greenhouse gas emissions;

Mitigation: Measures to slow down and prevent climate change, mainly GHG emissions reductions and energy efficiency improvements;

Adaptation: Measures to adjust to climatic changes, for example, building defenses against sea-level rise, growing different (e.g., more drought-resistant) crops, migrating;

Geo-engineering: Planned (engineered) large-scale interventions to change the climate, for example, cloud-whitening, air capture and storage of carbon dioxide, blocking (some) incoming sunlight by spraying small particles (aerosols) into the stratosphere.

IMPORTANT: If you want to read these instructions or the abbreviations and definitions a second time, press the help button at the bottom right corner.

Step 2: Ranking the Statements

Next, you will be shown a grid (–5 to +5) and the statements in the presorted piles at the bottom of the page. The goal is to place all the cards in the grid above based on your level of agreement or disagreement with them. What matters most is the final composition of the grid, not how you got there. Below I suggest one way to approach this ranking. If you choose another approach, please make sure that you spend some time in the end to move statements to their best place based on your beliefs (step 3).

Please read the statements in your "Agree" pile again. Select the *two* statements that you *agree with most* and place them on right side of the score sheet below the "+5." It does not matter which of these you place at the top or at the bottom. Next, from the remaining pile select the four statements that you *agree with most* and place them in the four spots below the "+4." Proceed until all statements you agree with have been placed on the grid.

Next, focus on the cards in the "Disagree" pile. Please read through the cards again and—just as before—select the two statements that you *disagree with most* and place them on the left side of the score sheet below the "–5." Proceed until all statements you disagree with have been placed on the grid.

Finally, look at the remaining cards in the "Neither/Conflicted" pile. Please read through these statements once again and place the cards in the remaining spots on the score sheet.

Please note that it is possible that you end up placing some statements that you agree with in the "0" column, or even in the "Disagree with" area of the grid, or vice versa. This is not a problem since the purpose of the grid is to establish the relative agreement with and importance of these statements. Please email me (manjana@mac.com) if you experience serious problems with this issue.

IMPORTANT: If you want to read this instruction a second time, press the help button at the bottom right corner.

Step 3: Check and Adjust Your Ranking
You have now placed all cards onto the grid. Please look over your grid once more for a final check and shift cards if you want to. You can drag statements out of their current location back into one of the three piles, and then relocate from there or swap cards directly.

Step 4: Help Us Interpret Your Ranking
Please explain why you have placed your selected statements under "+5" ("most agree with") and "–5" ("most disagree with").

Step 5: Some Final Questions
Almost done! The final step consists of a number of questions about yourself and about your ranking. This part might be the most valuable for

understanding your ranking and how your point of view compares with other. Therefore any brief comment would be greatly appreciated!

Questions 1–4 and 6–11 are required to finish the Q Sort; all other questions are voluntary.

Postsorting Questions

General Questions

1. Name
2. Nationality
3. Current Occupation
4. Country of Residence
5. Age

Participant Group Questions

1. Do you think climate change is an important problem—is it among the top five issues global leaders should worry about?
2. Is an international agreement necessary to address climate change?
3. Are you optimistic that climate change will be addressed effectively with cooperative international policies—is it a tractable problem?
4. Does your home country have comparatively high GHG emissions?
5. Do you expect your home country will be strongly, negatively affected by climate change?
6. Do you expect that you will be personally, negatively affected by climate change?

Interpretation Questions

1. Why did you place statements X and Y as "most agree with"?
2. Why did you place statements A and B under "most disagree with"?
3. Was a statement missing that you needed to represent your views on climate change and international cooperation?
4. Which statement did you struggle most with—which one was hardest to place?
5. Is there a statement in the "0" column that is important to you, but that you feel conflicted about, that you were not sure whether to place it in "agree with"/"disagree with"? Why?

Appendix A8   Q Method Factor Scores

Table A.4

Q method factor scores

Statement		Factor scores					
		A	B	C	D	E	F
1	Human-released greenhouse gases are causing significant climate change.	+3	+5	+5	+4	+5	+3
2	I don't trust what scientists say about climate change.	–2	–5	–5	–4	–4	–4
3	Anti–climate change policies threaten progress and modernity.	–3	–1	0	–1	–1	+3
4	I do not believe that we will see significant effects of climate change in my lifetime.	–4	–4	–3	–2	–3	–4
5	The social consequences of climate change are most worrying: Whole countries and cultures will disappear; many people will suffer from food and water scarcity, or will be forced to migrate.	+3	+2	0	+3	+3	+2
6	Climate change is mainly an issue for the developing countries.	–5	–4	–4	–5	–5	–3
7	Climate change was caused by rich, industrialized countries, but its impacts will hurt poor countries most—this asymmetry is unfair.	+1	+3	+1	+4	+5	+3
8	Climate change will result in violence and human deaths.	+1	+3	+1	+1	+2	+3
9	An effective climate solution requires the cooperation of all governments around the world.	+5	0	+4	+3	+3	+1
10	States are the most important players in global climate politics.	0	–1	+1	0	+1	+2
11	Even if a new international treaty is signed with ambitious mitigation targets, many states will simply not be able to implement these targets domestically.	+2	0	–1	0	–1	0
12	Only a small number of countries with significant GHG emissions are important for climate negotiations.	–4	+2	–3	–3	–1	–1
13	All our efforts and resources to combat climate change should be concentrated domestically, rather than internationally.	–4	–1	–4	–2	–2	–2
14	The power disparities between countries in climate negotiations make me very upset.	+1	0	–2	+2	+2	+4

Table A.4 (continued)

Statement	Factor scores					
	A	B	C	D	E	F
15 The climate problem should be left to the markets.	–3	–2	+1	–5	–4	–4
16 It is best to leave the development of climate solutions to regions, cities, and local communities—they have been much more successful than UNFCCC negotiations.	–1	–1	–2	–1	–1	–1
17 Individual contributions don't make a difference when it comes to climate change.	–1	–2	–2	–3	–2	–1
18 Politicians need much stronger voter support and pressure from political movements to create meaningful climate policies.	+3	+2	+4	+5	+4	+4
19 The vested interests blocking solutions are too powerful to allow for any meaningful action on climate change.	+2	+1	–2	+2	+2	+1
20 Nothing will happen before a climate crisis hits.	0	+1	–1	–1	–1	0
21 Neither states nor markets nor civil societies can solve this problem on their own—climate change is a multilevel problem and requires action at all of these different levels.	+4	+4	+4	+4	+4	+1
22 A key element in solving the climate problem is the need for fundamental value change within our societies.	+5	0	+3	+5	+1	+5
23 Based on our shared humanity, our desire for happiness and security, we can find a solution to the climate problem.	+1	+1	+2	+3	+2	0
24 God has made us stewards of the Earth, giving us the ability and responsibility to keep the planet healthy.	+2	–4	–2	0	+1	+1
25 The prospect of a cleaner, eco-friendly economy is exciting.	+3	+4	+4	+3	+2	+4
26 Limiting average global warming to 2°C will be sufficient to prevent major damage.	–3	–2	–1	–2	–2	–1
27 We lack a clearly defined goal for global climate policy.	0	0	0	+1	–1	–1
28 Since the climate is going to change we should be more concerned with adaptation.	+3	+1	–1	–1	+3	0
29 Adaptation and mitigation are complementary and equally important policies.	+4	+3	+2	+1	+2	+1

Table A.4 (continued)

Statement	A	B	C	D	E	F
	Factor scores					
30 Given the political gridlock on mitigation, we might have to resort to geo-engineering to buy more time for mitigation and adaptation.	–2	0	0	–2	0	0
31 Geo-engineering can solve the global warming problem much more cost-effectively than mitigation.	–1	–3	–2	–3	–2	–3
32 Taxes—whether globally or domestically—and other policies that constrain private property rights are simply not politically acceptable in our system.	0	–1	–3	–2	–3	–3
33 Economic growth and jobs must take priority over climate concerns.	–2	–3	–2	–3	–2	–2
34 Ideally, climate policies would reduce GHG emissions while stimulating economic growth.	+2	+4	+3	+1	+3	+1
35 Climate change policy should be based on cost-benefit analysis.	–3	0	+2	–2	–1	+1
36 The main costs of climate change policies include loss of GDP and jobs.	–3	–1	0	–1	–2	–2
37 The main costs of future climate change can simply not be calculated: the loss of human life, food insecurity, or species extinctions don't have price tags.	+1	+2	+2	+2	+1	–2
38 Economic growth is the best solution to climate change.	–5	–2	+1	–3	–3	–2
39 Investment in climate policies is a poor use of our resources; it makes more sense to do something about poverty, health care, and education in the developing world.	–2	–3	–5	–2	–3	–2
40 The prospect of major environmental change, such as the melting of glaciers or species extinction, is very distressing.	+1	+3	+2	+2	+3	+1
41 I am ashamed that my country is not doing more about climate change.	–2	+2	–3	+1	+1	+4
42 The BASIC countries (Brazil, South Africa, India, and China) should show greater leadership in international climate negotiations.	+4	+1	+3	+1	+1	–1
43 I dedicate my career to solving the climate problem.	–1	+2	+3	0	0	0

Table A.4 (continued)

Statement	A	B	C	D	E	F
		Factor scores				
44 It is disappointing to see how little governments and markets care about the environment and the health of the planet.	+2	+1	0	+3	+1	+2
45 My government is a very constructive player in international affairs.	0	−3	+2	−1	−3	−5
46 Elected officials have a political responsibility to protect the interests of their constituency—the present rather than future generations.	−1	−3	−1	−3	−2	−3
47 Problems that might arise decades from now are not important to me.	−4	−5	−4	−4	−4	−5
48 All states have a moral responsibility to contribute to a global climate solution.	+4	+3	+5	+2	+3	+3
49 The current generation (of politicians and voters) has a major ethical responsibility to future generations.	+2	+4	+3	+4	+4	+5
50 Future generations are likely to be richer and better off than we are, and better able to deal with climate change.	−2	−2	−1	−4	−3	−4
51 I fully support climate funding—financial flows from the rich to the poor to help them cope with climate change.	+3	+3	0	+3	+2	+3
52 The rich countries have caused the problem; consequently, they have the obligation to fix it.	−2	0	−1	+1	+4	+2
53 Contemplating climate change usually leaves me feeling rather helpless.	−1	−2	−3	−1	−1	+1
54 Climate change makes me fear for my children's future.	0	+2	+1	+2	0	0
55 I believe that we will find a cooperative solution to climate change. Other issues have taken many years of negotiation, too.	+1	+1	+3	+1	+1	0
56 Climate change is a very depressing issue.	0	0	0	0	0	+2
57 Climate change is simply too complex and overwhelming. It is impossible to fully understand, let alone solve, the problem.	−1	−3	−3	−1	−4	−3
58 It is already too late to do anything about climate change.	−3	−4	−4	−4	−5	−3
59 There are moments when I lose all hope that the UNFCCC process can solve this problem.	+1	+1	0	0	0	+2
60 Climate change scares me because I don't know what's going to happen.	−1	−1	−1	0	−1	0

Table A.4 (continued)

Statement	Factor scores					
	A	B	C	D	E	F
61 Climate change is not the only issue we have to deal with and other issues are often more urgent.	+1	–2	+1	0	0	–2
62 The focus on winning the next election is the biggest obstacle to finding international agreement.	–1	+1	+1	–1	0	–1
63 I sometimes wonder how to explain our failure to fix the climate problem to my grandchildren.	0	–1	+1	+1	0	–1
64 I have a hard time imagining the consequences of climate change for my community and my country.	0	–1	–1	0	0	–1
65 Sometimes gradual processes such as GHG emissions result in sudden, dramatic changes in the environment. The existence of such climate tipping points makes action even more urgent than previously thought. Avoiding tipping points should become a key climate policy goal.	+2	+5	+2	+2	+1	+2

References

Abdelal, Rawi, Yoshiko M. Herrera, Alastair Iain Johnston, and Rose McDermott. 2006. Identity as a Variable. *Perspectives on Politics* 4 (4): 695–711. doi:10.1017/S1537592706060440.

Adger, W. Neil. 2006. Vulnerability. *Global Environmental Change* 16 (3): 268–281. doi:10.1016/j.gloenvcha.2006.02.006.

Adger, W. Neil, Nigel W. Arnell, and Emma L. Tompkins. 2005. Successful Adaptation to Climate Change across Scales. *Global Environmental Change* 15 (2): 77–86. doi:10.1016/j.gloenvcha.2004.12.005.

Adger, W. Neil, Jon Barnett, F. S. Chapin, and Heidi Ellemor. 2011. This Must Be the Place: Underrepresentation of Identity and Meaning in Climate Change Decision-Making. *Global Environmental Politics* 11 (2): 1–25. doi:10.1162/GLEP_a_00051.

Adger, W. Neil, Suraje Dessai, Marisa Goulden, Mike Hulme, Irene Lorenzoni, Donald Nelson, et al. 2009. Are There Social Limits to Adaptation to Climate Change? *Climatic Change* 93 (3): 335–354. doi:10.1007/s10584-008-9520-z.

Akhtar-Danesh, Noori, Andrea Baumann, and Lis Cordingley. 2008. Q-Methodology in Nursing Research: A Promising Method for the Study of Subjectivity. *Western Journal of Nursing Research* 30 (6): 759–773. doi:10.1177/0193945907312979.

Albin, Cecilia. 2001. *Justice and Fairness in International Negotiation*. Cambridge: Cambridge University Press.

Alexander, Michele G., Shana Levin, and P. J. Henry. 2005. Image Theory, Social Identity, and Social Dominance: Structural Characteristics and Individual Motives Underlying International Images. *Political Psychology* 26 (1): 27–45. doi:10.1111/j.1467-9221.2005.00408.x.

Anderson, Kevin. 2015. Duality in Climate Science. *Nature Geoscience* 8 (12): 898–900. doi:10.1038/ngeo2559.

Anderson, Kevin, and Alice Bows. 2012. A New Paradigm for Climate Change. *Nature Climate Change* 2 (9): 639–640. doi:10.1038/nclimate1646.

Andonova, Liliana B., Michele M. Betsill, and Harriet Bulkeley. 2011. Transnational Climate Governance. *Global Environmental Politics* 9 (2): 52–73. doi:10.1162/glep.2009.9.2.52

Antal, Miklós, and Janne I. Hukkinen. 2010. The Art of the Cognitive War to Save the Planet. *Ecological Economics* 69 (5): 937–943.

Antilla, Liisa. 2005. Climate of Scepticism: U.S. Newspaper Coverage of the Science of Climate Change. *Global Environmental Change* 15 (4): 338–352. doi:10.1016/j.gloenvcha.2005.08.003.

Armitage, Derek. 2005. Adaptive Capacity and Community-Based Natural Resource Management. *Environmental Management* 35 (6): 703–715. doi:10.1007/s00267-004-0076-z.

Astorino-Courtois, Allison. 1995. The Cognitive Structure of Decision Making and the Course of Arab-Israeli Relations, 1970–1978. *Journal of Conflict Resolution* 39 (3): 419–438.

Atran, Scott, and Robert Axelrod. 2008. Reframing Sacred Values. *Negotiation Journal* 24 (3): 221–246. doi:10.1111/j.1571-9979.2008.00182.x.

Atran, Scott, and Jeremy Ginges. 2012. Religious and Sacred Imperatives in Human Conflict. *Science* 336 (6083): 855–857. doi:10.1126/science.1216902.

Axelrod, Robert M. 1976. *Structure of Decision: The Cognitive Maps of Political Elites.* Princeton: Princeton University Press.

Bain, Paul G., Matthew J. Hornsey, Renata Bongiorno, and Carla Jeffries. 2012. Promoting Pro-Environmental Action in Climate Change Deniers. *Nature Climate Change* 2 (8): 600–603. doi:10.1038/nclimate1532.

Bandura, Albert. 1989. Human Agency in Social Cognitive Theory. *American Psychologist* 44 (9): 1175–1184. doi:10.1037/0003-066X.44.9.1175.

Bandura, Albert. 2001. Social Cognitive Theory: An Agentic Perspective. *Annual Review of Psychology* 52 (1): 1–26. doi:10.1146/annurev.psych.52.1.1.

Bandura, Albert. 2006. Toward a Psychology of Human Agency. *Perspectives on Psychological Science* 1 (2): 164–180. doi:10.1111/j.1745-6916.2006.00011.x.

Barrett, Scott. 2007. *Why Cooperate? The Incentive to Supply Global Public Goods.* Oxford: Oxford University Press.

Barrett, Scott, and Astrid Dannenberg. 2012. Climate Negotiations under Scientific Uncertainty. *Proceedings of the National Academy of Sciences of the United States of America* 109 (43): 17372–17376. doi:10.1073/pnas.1208417109.

Barrett, Scott, and Astrid Dannenberg. 2014. Sensitivity of Collective Action to Uncertainty about Climate Tipping Points. *Nature Climate Change* 4 (1): 36–39. doi:10.1038/nclimate2059.

Barry, John, and John Proops. 1999. Seeking Sustainability Discourses with Q Methodology. *Ecological Economics* 28 (3): 337–345. doi:10.1016/S0921-8009(98)00053-6.

Bar-Tal, Daniel. 2001. Why Does Fear Override Hope in Societies Engulfed by Intractable Conflict, as It Does in the Israeli Society? *Political Psychology* 22 (3): 601–627. doi:10.1111/0162-895X.00255.

Bellamy, Rob, and Mike Hulme. 2011. Beyond the Tipping Point: Understanding Perceptions of Abrupt Climate Change and Their Implications. *Weather, Climate, and Society* 3 (1): 48–60. doi:10.1175/2011WCAS1081.1.

Benford, Robert D., and David A. Snow. 2000. Framing Processes and Social Movements: An Overview and Assessment. *Annual Review of Sociology* 26 (January): 611–639.

Bernauer, Thomas. 2013. Climate Change Politics. *Annual Review of Political Science* 16 (1): 421–448. doi:10.1146/annurev-polisci-062011-154926.

Berns, Gregory S., and Scott Atran. 2012. The Biology of Cultural Conflict. *Philosophical Transactions of the Royal Society of London. Series B, Biological Sciences* 367 (1589): 633–639. doi:10.1098/rstb.2011.0307.

Berns, Gregory S., Emily Bell, C. Monica Capra, Michael J. Prietula, Sara Moore, Brittany Anderson, et al. 2012. The Price of Your Soul: Neural Evidence for the Non-Utilitarian Representation of Sacred Values. *Philosophical Transactions of the Royal Society of London, Series B, Biological Sciences* 367 (1589): 754–762. doi:10.1098/rstb.2011.0262.

Bernstein, Steven F. 2001. *The Compromise of Liberal Environmentalism*. New York: Columbia University Press.

Betsill, Michele M, and Harriet Bulkeley. 2006. Cities and the Multilevel Governance of Global Climate Change. *Global Governance* 12:141.

Betsill, Michele M., and Elisabeth Corell. 2011. NGO Influence in International Environmental Negotiations: A Framework for Analysis. *Global Environmental Politics* 1 (4): 65–85. doi:10.1162/152638001317146372.

Biermann, Frank. 2014. The Anthropocene: A Governance Perspective. *Anthropocene Review* 1 (1): 57–61. doi:10.1177/2053019613516289.

Bischof, Bärbel G. 2010. Negotiating Uncertainty: Framing Attitudes, Prioritizing Issues, and Finding Consensus in the Coral Reef Environment Management "Crisis." *Ocean and Coastal Management* 53 (10): 597–614. doi:10.1016/j.ocecoaman.2010.06.020.

Blackstock, K. L., G. J. Kelly, and B. L. Horsey. 2007. Developing and Applying a Framework to Evaluate Participatory Research for Sustainability. *Ecological Economics* 60 (4): 726–742. doi:10.1016/j.ecolecon.2006.05.014.

Blake, James. 1999. Overcoming the "Value-Action Gap" in Environmental Policy: Tensions between National Policy and Local Experience. *Local Environment* 4 (3): 257.

Bleiker, Roland, and Emma Hutchison. 2008. Fear No More: Emotions and World Politics. *Review of International Studies* 34 (Supplement S1): 115–135. doi:10.1017/S0260210508007821.

Bleiker, Roland, and Emma Hutchison. 2014. Introduction: Emotions and World Politics. *International Theory* 6 (3): 490–491. doi:10.1017/S1752971914000220.

Bodansky, Daniel. 2011. A Tale of Two Architectures: The Once and Future U.N. Climate Change Regime. *SSRN eLibrary*, March. http://papers.ssrn.com/sol3/papers.cfm?abstract_id=1773865.

Bodansky, Daniel. 2016. The Legal Character of the Paris Agreement. SSRN Scholarly Paper ID 2735252. Rochester, NY: Social Science Research Network. http://papers.ssrn.com/abstract=2735252.

Bonham, G. 1993. Cognitive Mapping as a Technique for Supporting International Negotiation. *Theory and Decision* 34 (3): 255–273.

Bonham, G. Matthew, Michael J. Shapiro, and Daniel Heradstveit. 1988. Group Cognition: Using an Oil Policy Game to Validate a Computer Simulation. *Simulation & Games* 19:379–407.

Brechin, Steven R., and Medani Bhandari. 2011. Perceptions of Climate Change Worldwide. *Wiley Interdisciplinary Reviews: Climate Change* 2 (6): 871–885. doi:10.1002/wcc.146.

Brenton, Anthony. 2013. "Great Powers" in Climate Politics. *Climate Policy* 13 (5): 541–546. doi:10.1080/14693062.2013.774632.

Broecker, Wallace S. 1987. Unpleasant Surprises in the Greenhouse? *Nature* 328 (6126): 123–126. doi:10.1038/328123a0.

Brown, Steven R. 1980. *Political Subjectivity: Applications of Q Methodology in Political Science*. New Haven: Yale University Press.

Brown, Steven R., and Richard Robyn. 2004. Reserving a Key Place for Reality: Philosophical Foundations of Theoretical Rotation. *Operant Subjectivity* 27 (3): 104–124. doi:10.15133/j.os.2004.004.

Bruneau, Emile G., Nicholas Dufour, and Rebecca Saxe. 2012. Social Cognition in Members of Conflict Groups: Behavioural and Neural Responses in Arabs, Israelis and South Americans to Each Other's Misfortunes. *Philosophical Transactions of the Royal Society of London. Series B, Biological Sciences* 367 (1589): 717–730. doi:10.1098/rstb.2011.0293.

Bulkeley, Harriet, Liliana Andonova, Michele M. Betsill, Daniel Compagnon, Thomas Hale, Matthew J. Hoffmann, et al. 2014. *Transnational Climate Change Governance*. Cambridge: Cambridge University Press.

Bulkeley, Harriet, and Peter Newell. 2015. *Governing Climate Change*. London: Routledge.

Checkel, Jeffrey T. 1998. The Constructivist Turn in International Relations Theory. *World Politics* 50 (2): 324–348.

Clark, Peter U., Jeremy D. Shakun, Shaun A. Marcott, Alan C. Mix, Michael Eby, Scott Kulp, et al. 2016. Consequences of Twenty-First-Century Policy for Multi-Millennial Climate and Sea-Level Change. *Nature Climate Change* (February): advance online publication. doi:10.1038/nclimate2923.

Collins, Kevin, and Ray Ison. 2009. Jumping off Arnstein's Ladder: Social Learning as a New Policy Paradigm for Climate Change Adaptation. *Environmental Policy and Governance* 19 (6): 358–373. doi:10.1002/eet.523.

Connolly, William E. 2002. *Neuropolitics: Thinking, Culture, Speed*. Minneapolis: University of Minnesota Press.

Cottam, Martha L. 1986. *Foreign Policy Decision Making: The Influence of Cognition*. Boulder, CO: Westview Press.

Courville, Sasha, and Nicola Piper. 2004. Harnessing Hope through NGO Activism. *Annals of the American Academy of Political and Social Science* 592 (1): 39–61. doi:10.1177/0002716203261940.

Cox, Robert W. 1996. *Approaches to World Order*. Cambridge: Cambridge University Press.

Crawford, Neta C. 2000. The Passion of World Politics: Propositions on Emotion and Emotional Relationships. *International Security* 24 (4): 116–156.

Crawford, Neta C. 2014. Institutionalizing Passion in World Politics: Fear and Empathy. *International Theory* 6 (3): 535–557. doi:10.1017/S1752971914000256.

Crona, Beatrice, Amber Wutich, Alexandra Brewis, and Meredith Gartin. 2013. Perceptions of Climate Change: Linking Local and Global Perceptions through a Cultural Knowledge Approach. *Climatic Change* 119 (2): 519–531. doi:10.1007/s10584-013-0708-5.

Damasio, Antonio. 1995. *Descartes' Error: Emotion, Reason, and the Human Brain*. New York: Harper Perennial.

Damasio, Antonio. 2003. *Looking for Spinoza: Joy, Sorrow, and the Feeling Brain*. Orlando, FL: Harvest.

Demertzis, Nicolas. 2013. *Emotions in Politics: The Affect Dimension in Political Tension*. New York: Palgrave Macmillan.

Depledge, Joanna. 2006. The Opposite of Learning: Ossification in the Climate Change Regime. *Global Environmental Politics* 6 (1): 1–22. doi:10.1162/glep.2006.6.1.1.

Dessai, Suraje, W. Neil Adger, Mike Hulme, John Turnpenny, Jonathan Köhler, and Rachel Warren. 2004. Defining and Experiencing Dangerous Climate Change. *Climatic Change* 64 (1/2): 11–25. doi:10.1023/B:CLIM.0000024781.48904.45.

Dirlik, Arif. 1999. Place-Based Imagination: Globalism and the Politics of Place. *Review: A Journal of the Fernand Braudel Center* 22 (2): 151–187.

Douglas, Mary. 1986. *How Institutions Think*. Syracuse: Syracuse University Press.

Douglas, Mary, and Aaron Wildavsky. 1982. *Risk and Culture: An Essay on the Selection of Technological and Environmental Dangers*. Berkeley: University of California Press.

Dryzek, John S. 2014. Institutions for the Anthropocene: Governance in a Changing Earth System. *British Journal of Political Science FirstView* Supplement-1:1–20. doi:10.1017/S0007123414000453.

Dryzek, John S., and Jeffrey Berejikian. 1993. Reconstructive Democratic Theory. *American Political Science Review* 87 (1): 48. doi:10.2307/2938955.

Dryzek, John S., Richard B. Norgaard, and David Schlosberg, eds. 2011. *The Oxford Handbook of Climate Change and Society*. New York: Oxford University Press.

Duncan, Seth, and Lisa Feldman Barrett. 2007. Affect Is a Form of Cognition: A Neurobiological Analysis. *Cognition and Emotion* 21 (6): 1184–1211. doi:10.1080/02699930701437931.

Eckersley, Robyn. 2012. Moving Forward in the Climate Negotiations: Multilateralism or Minilateralism? *Global Environmental Politics* 12 (2): 24–42. doi:10.1162/GLEP_a_00107.

Eliasmith, Chris, and Charles H. Anderson. 2004. *Neural Engineering: Computation, Representation, and Dynamics in Neurobiological Systems*. New ed. Cambridge, MA: MIT Press.

Ellemers, Naomi. 2012. The Group Self. *Science* 336 (6083): 848–852. doi:10.1126/science.1220987.

Ellemers, Naomi, and Nick S. Alexander Haslam. 2012. Social Identity Theory. In *Handbook of Theories of Social Psychology*, Vol. 2, ed. Paul A. M. Van Lange, Arie W. Kruglanski, and E. Tory Higgins, 379–398. Thousand Oaks, CA: Sage.

Epstein, Seymour. 1994. Integration of the Cognitive and the Psychodynamic Unconscious. *American Psychologist* 49 (8): 709–724. doi:10.1037/0003-066X .49.8.709.

Falkner, Robert. 2005. American Hegemony and the Global Environment. *International Studies Review* 7 (4): 585–599. doi:10.1111/j.1468-2486.2005.00534.x.

Falkner, Robert. 2006. International Cooperation against the Hegemon: The Cartagena Protocol on Biosafety. In *The International Politics of Genetically Modified Food*: 15–33. Palgrave Macmillan UK.

Falkner, Robert. 2008. *Business Power and Conflict in International Environmental Politics*. Houndmills, England: Palgrave Macmillan.

Feinberg, Matthew, and Robb Willer. 2013. The Moral Roots of Environmental Attitudes. *Psychological Science* 24 (1): 56–62. doi:10.1177/0956797612449177.

Feldman, Stanley. 1988. Structure and Consistency in Public Opinion: The Role of Core Beliefs and Values. *American Journal of Political Science* 32 (2): 416. doi:10.2307/2111130.

Findlay, Scott, and Paul Thagard. 2012. Emotional Change in International Negotiation: Analyzing the Camp David Accords Using Cognitive-Affective Maps. *Group Decision and Negotiation* 23 (6): 1–20. Accessed August 17. doi:10.1007/ s10726-011-9242-x.

Finucane, Melissa L., Ali Alhakami, Paul Slovic, and Stephen M. Johnson. 2000. The Affect Heuristic in Judgments of Risks and Benefits. *Journal of Behavioral Decision Making* 13 (1): 1–17. doi:10.1002/(SICI)1099-0771(200001/03)13:1<1:AID-BDM333> 3.0.CO;2-S.

Frankl, Viktor Emil. 1959. *Man's Search for Meaning: An Introduction to Logotherapy*. New York: Simon & Schuster.

Fresque-Baxter, Jennifer A., and Derek Armitage. 2012. Place Identity and Climate Change Adaptation: A Synthesis and Framework for Understanding. *Wiley Interdisciplinary Reviews: Climate Change* 3 (3): 251–266. doi:10.1002/wcc.164.

Funtowicz, Silvio O., and Jerome R. Ravetz. 1993. Science for the Post-Normal Age. *Futures* 25 (7): 739–755. doi:10.1016/0016-3287(93)90022-L.

Galushkin, Alexander I. 2007. *Neural Networks Theory*. Berlin: Springer.

Gampfer, Robert. 2014. Do Individuals Care about Fairness in Burden Sharing for Climate Change Mitigation? Evidence from a Lab Experiment. *Climatic Change* 124 (1–2): 65–77. doi:10.1007/s10584-014-1091-6.

Gamson, William A. 1999. Beyond the Science-versus-Advocacy Distinction. *Contemporary Sociology* 28 (1): 23–26.

Gardiner, Stephen M. 2009. Saved by Disaster? Abrupt Climate Change, Political Inertia, and the Possibility of an Intergenerational Arms Race. *Journal of Social Philosophy* 40 (2): 140–162.

Gardiner, Stephen M. 2010. Ethics and Climate Change: An Introduction. *Wiley Interdisciplinary Reviews: Climate Change* 1 (1): 54–66. doi:10.1002/wcc.16.

Gardiner, Stephen M. 2011. *A Perfect Moral Storm: The Ethical Tragedy of Climate Change*. New York: Oxford University Press.

Geden, Oliver, and Silke Beck. 2014. Renegotiating the Global Climate Stabilization Target. *Nature Climate Change* 4 (9): 747–748. doi:10.1038/nclimate2309.

George, Alexander L. 1969. The "Operational Code": A Neglected Approach to the Study of Political Leaders and Decision-Making. *International Studies Quarterly* 13 (2): 190–222.

Giddens, Anthony. 1991. *Modernity and Self-Identity: Self and Society in the Late Modern Age*. Stanford: Stanford University Press.

Giddens, Anthony. 1992. *The Constitution of Society: Outline of the Theory of Structuration*. Berkeley: University of California Press.

Gidley, Jennifer M., John Fien, Jodi-Anne Smith, Dana C. Thomsen, and Timothy F. Smith. 2009. Participatory Futures Methods: Towards Adaptability and Resilience in Climate-Vulnerable Communities. *Environmental Policy and Governance* 19 (6): 427–440. doi:10.1002/eet.524.

Gifford, Robert. 2011. The Dragons of Inaction: Psychological Barriers That Limit Climate Change Mitigation and Adaptation. *American Psychologist* 66 (4): 290–302. doi:10.1037/a0023566.

Ginges, Jeremy, and Scott Atran. 2011. War as a Moral Imperative (Not Just Practical Politics by Other Means). *Proceedings of the Royal Society, Series B, Biological Sciences* 278 (1720): 2930–2938. doi:10.1098/rspb.2010.2384.

Gladwell, Malcolm. 2002. *The Tipping Point: How Little Things Can Make a Big Difference*. Boston: Back Bay.

Gollier, Christian, and Martin L. Weitzman. 2010. How Should the Distant Future Be Discounted When Discount Rates Are Uncertain? *Economics Letters* 107 (3): 350–353. doi:10.1016/j.econlet.2010.03.001.

Grasso, Marco. 2013. Climate Ethics: With a Little Help from Moral Cognitive Neuroscience. *Environmental Politics* 22 (3): 377–393. doi:10.1080/09644016.2012.730263.

Grasso, Marco, and Ezra M. Markowitz. 2015. The Moral Complexity of Climate Change and the Need for a Multidisciplinary Perspective on Climate Ethics. *Climatic Change* 130 (3): 327–334. doi:10.1007/s10584-014-1323-9.

Greene, Joshua D. 2008. The Secret Joke of Kant's Soul. In *Moral Psychology*, Vol. 3, *The Neuroscience of Morality: Emotion, Brain Disorders, and Development*, ed. Walter Sinnott-Armstrong, 35–80. Cambridge, MA: MIT Press.

Gromet, Dena M., Howard Kunreuther, and Richard P. Larrick. 2013. Political Ideology Affects Energy-Efficiency Attitudes and Choices. *Proceedings of the National Academy of Sciences of the United States of America* 110 (23): 9314–9319. doi:10.1073/pnas.1218453110.

Gross Stein, Janice, and David Welch. 1997. Rational and Psychological Approaches to the Study of International Conflict: Comparative Strengths and Weaknesses. In *Decisionmaking on War and Peace: The Cognitive Rational Debate*, ed. Nehemia Geva and Alex Mintz, 51–77. Boulder, CO: Lynne Rienner.

Gruber, James S. 2011. Perspectives of Effective and Sustainable Community-Based Natural Resource Management: An Application of Q Methodology to Forest Projects. *Conservation & Society* 9 (2): 159.

Grundig, Frank. 2006. Patterns of International Cooperation and the Explanatory Power of Relative Gains: An Analysis of Cooperation on Global Climate Change, Ozone Depletion, and International Trade. *International Studies Quarterly* 50 (4): 781–801. doi:10.1111/j.1468-2478.2006.00425.x.

Gunnell, John G. 2007. Are We Losing Our Minds? Cognitive Science and the Study of Politics. *Political Theory* 35 (6): 704–731. doi:10.1177/0090591707307327.

Guston, David H. 2001. Boundary Organizations in Environmental Policy and Science: An Introduction. *Science, Technology & Human Values* 26 (4): 399–408.

Haidt, Jonathan. 2001. The Emotional Dog and Its Rational Tail: A Social Intuitionist Approach to Moral Judgment. *Psychological Review* 108 (4): 814–834. doi:10.1037/0033-295X.108.4.814.

Haidt, Jonathan. 2003. The Moral Emotions. In *Handbook of Affective Sciences*, ed. Richard J. Davidson, Klaus R. Scherer, and H. Hill Goldsmith, 852–870. Oxford: Oxford University Press.

Haidt, Jonathan. 2013. *The Righteous Mind: Why Good People Are Divided by Politics and Religion*. New York: Random House.

Hajer, Maarten A. 1996. *Politics of Environmental Discourse: Ecological Modernization and the Policy Process*. New York: Oxford University Press.

Hall, Rodney Bruce. 1999. *National Collective Identity: Social Constructs and International Systems*. New York: Columbia University Press.

Hallding, Karl, Marie Jürisoo, Marcus Carson, and Aaron Atteridge. 2013. Rising Powers: The Evolving Role of BASIC Countries. *Climate Policy* 13 (5): 608–631. doi:10.1080/14693062.2013.822654.

Hamm, Robert M., Michelle A. Miller, and Richard S. Ling. 1992. Preferences, Beliefs, and Values in Negotiations Concerning Aid to Nicaragua. *Public Choice* 74 (1): 79–103.

Hampson, Fen Osler. 1989. Climate Change: Building International Coalitions of the Like-Minded. *International Journal (Toronto)* 45 (1): 36–74. doi:10.2307/40202651.

Hanemann, W. Michael. 2008. What Is the Economic Cost of Climate Change? Berkeley: University of California, Department of Agricultural and Resource Economics, August. http://escholarship.org/uc/item/9g11z5cc.

Hanselmann, Martin, and Carmen Tanner. 2008. Taboos and Conflicts in Decision Making: Sacred Values, Decision Difficulty, and Emotions. *Judgment and Decision Making* 3 (1): 51.

Hansen, J., M. Sato, P. Hearty, R. Ruedy, M. Kelley, V. Masson-Delmotte, et al. 2015. Ice Melt, Sea Level Rise and Superstorms: Evidence from Paleoclimate Data, Climate Modeling, and Modern Observations That 2°C Global Warming Is Highly Dangerous. *Atmospheric Chemistry and Physics Discussion* 15 (14): 20059–20179. doi:10.5194/acpd-15-20059-2015.

Hardin, Garrett. 1968. The Tragedy of the Commons. *Science* 162 (3859): 1243–1248. doi:10.1126/science.162.3859.1243.

Harré, Rom, and Grant Gillett. 1994. *The Discursive Mind*. Atlanta: Sage.

Harrison, Kathryn, and Lisa McIntosh Sundstrom. 2010. *Global Commons, Domestic Decisions: The Comparative Politics of Climate Change*. Cambridge, MA: MIT Press.

Haslam, S. A., and N. Ellemers. 2016. Social identification is generally a prerequisite for group success and does not preclude intragroup differentiation. *Behavioral and Brain Sciences* 39: 150.

Hatemi, Peter K, and Rose McDermott. 2011. *Man Is by Nature a Political Animal: Evolution, Biology, and Politics*. Chicago: University of Chicago Press.

Hegger, Dries, Machiel Lamers, Annemarie Van Zeijl-Rozema, and Carel Dieperink. 2012. Conceptualising Joint Knowledge Production in Regional Climate Change Adaptation Projects: Success Conditions and Levers for Action. *Environmental Science & Policy* 18 (April): 52–65. doi:10.1016/j.envsci.2012.01.002.

Held, David. 1995. *Democracy and the Global Order: From the Modern State to Cosmopolitan Governance*. Stanford: Stanford University Press.

Herrmann, Richard K., James F. Voss, Tonya Y. E. Schooler, and Joseph Ciarrochi. 1997. Images in International Relations: An Experimental Test of Cognitive Schemata. *International Studies Quarterly* 41 (3): 403–433.

Heymann, Matthias. 2010. The Evolution of Climate Ideas and Knowledge. *Wiley Interdisciplinary Reviews: Climate Change* 1 (4): 581–597. doi:10.1002/wcc.61.

Hochstetler, Kathryn, and Manjana Milkoreit. 2014. Emerging Powers in the Climate Negotiations Shifting Identity Conceptions. *Political Research Quarterly* 67 (1): 224–235. doi:10.1177/1065912913510609.

Hochstetler, Kathryn, and Eduardo Viola. 2012. Brazil and the Politics of Climate Change: Beyond the Global Commons. *Environmental Politics* 21 (5): 753–771. doi: 10.1080/09644016.2012.698884.

Hoffmann, Matthew J. 2011. *Climate Governance at the Crossroads: Experimenting with a Global Response after Kyoto.* Oxford: Oxford University Press.

Homer-Dixon, Thomas, Manjana Milkoreit, Steven J. Mock, Tobias Schröder, and Paul Thagard. 2014. The Conceptual Structure of Social Disputes. *SAGE Open* 4 (1). doi:10.1177/2158244014526210.

Homer-Dixon, Thomas, Brian Walker, Reinette Biggs, Anne-Sophie Crépin, Carl Folke, Eric F. Lambin, et al. 2015. Synchronous Failure: The Emerging Causal Architecture of Global Crisis. *Ecology and Society* 20 (3). doi:10.5751/ES-07681-200306.

Hulme, Mike. 2009. *Why We Disagree about Climate Change: Understanding Controversy, Inaction and Opportunity.* Cambridge: Cambridge University Press.

Hulme, Mike, and Martin Mahony. 2010. Climate Change: What Do We Know about the IPCC? *Progress in Physical Geography* 34 (5): 705–718. doi:10.1177/0309133310373719.

Hurrell, Andrew, and Sandeep Sengupta. 2012. Emerging Powers, North–South Relations and Global Climate Politics. *International Affairs* 88 (3): 463–484. doi:10.1111/j.1468-2346.2012.01084.x.

Hutchison, Emma, and Roland Bleiker. 2014. Theorizing Emotions in World Politics. *International Theory* 6 (3): 491–514. doi:10.1017/S1752971914000232.

Inhofe, Senator James. 2012. *The Greatest Hoax: How the Global Warming Conspiracy Threatens Your Future.* Washington, DC: WND Books.

IPCC (Intergovernmental Panel on Climate Change). 2001. *Climate Change 2001: Impacts, Adaptation, and Vulnerability.* Cambridge: Cambridge University Press. http://www.grida.no/publications/other/ipcc_tar/.

IPCC (Intergovernmental Panel on Climate Change). 2014. *Climate Change 2014: Mitigation of Climate Change. Contribution of Working Group III to the Fifth Assessment Report of the Intergovernmental Panel on Climate Change.* [Edenhofer, O., R. Pichs-Madruga, Y. Sokona, E. Farahani, S. Kadner, K. Seyboth, A. Adler, I. Baum, S. Brunner, P. Eickemeier, B. Kriemann, J. Savolainen, S. Schlömer, C. von Stechow, T. Zwickel and J.C. Minx (eds.)] Cambridge: Cambridge University Press.

Jacobs, Alan M., and J. Scott Matthews. 2012. Why Do Citizens Discount the Future? Public Opinion and the Timing of Policy Consequences. *British Journal of Political Science* 42 (4): 903–935. doi:10.1017/S0007123412000117.

Jacques, Peter J. 2012. A General Theory of Climate Denial. *Global Environmental Politics* 12 (2): 9–17. doi:10.1162/GLEP_a_00105.

Jacques, Peter J., Riley E. Dunlap, and Mark Freeman. 2008. The Organisation of Denial: Conservative Think Tanks and Environmental Scepticism. *Environmental Politics* 17 (3): 349. doi:10.1080/09644010802055576.

Jaeger, Carlo C., and Julia Jaeger. 2010. Three Views of Two Degrees. *Regional Environmental Change* 11 (1): 15–26. doi:10.1007/s10113-010-0190-9.

Jamison, Andrew. 2010. Climate Change Knowledge and Social Movement Theory. *Wiley Interdisciplinary Reviews: Climate Change* 1 (6): 811–823. doi:10.1002/wcc.88.

Jasanoff, Sheila. 2004. *States of Knowledge: The Co-Production of Science and the Social Order.* London: Routledge.

Jones, Natalie A., Helen Ross, Timothy Lynam, and Pascal Perez. 2014. Eliciting Mental Models: A Comparison of Interview Procedures in the Context of Natural Resource Management. *Ecology and Society* 19 (1). doi:10.5751/ES-06248-190113.

Jones, Natalie A., Helen Ross, Timothy Lynam, Pascal Perez, and Anne Leitch. 2011. Mental Models: An Interdisciplinary Synthesis of Theory and Methods. *Ecology and Society* 16 (12): 46.

Jordan, Andrew, Tim Rayner, Heike Schroeder, Neil Adger, Kevin Anderson, Alice Bows, et al. 2013. Going beyond Two Degrees? The Risks and Opportunities of Alternative Options. *Climate Policy* 13 (6): 751–769. doi:10.1080/14693062.2013.835 705.

Jost, John T., Brian A. Nosek, and Samuel D. Gosling. 2008. Ideology: Its Resurgence in Social, Personality, and Political Psychology. *Perspectives on Psychological Science* 3 (2): 126–136. doi:10.1111/j.1745-6916.2008.00070.x.

Kaarbo, Juliet. 2003. Foreign Policy Analysis in the Twenty-First Century: Back to Comparison, Forward to Identity and Ideas. *International Studies Review* 5 (2): 155–202.

Kahan, Dan. 2012. Cultural Cognition as a Conception of the Cultural Theory of Risk. In *Handbook of Risk Theory*, ed. Sabine Roeser, Rafaela Hillerbrand, Per Sandin, and Martin Peterson, 725–759. Dordrecht: Springer Netherlands. http://link.springer .com/referenceworkentry/10.1007/978-94-007-1433-5_28.

Kahan, Dan, Hank Jenkins-Smith, and Donald Braman. 2011. Cultural Cognition of Scientific Consensus. *Journal of Risk Research* 14 (2): 147–174. doi:10.1080/13669877 .2010.511246.

Kahneman, Daniel. 2011. *Thinking, Fast and Slow*. New York: Farrar, Straus and Giroux.

Kahneman, Daniel, and Amos Tversky. 1979. Prospect Theory: An Analysis of Decision under Risk. *Econometrica* 47 (2): 263–291. doi:10.2307/1914185.

Kaplan, Stanley, and B. John Garrick. 1981. On the Quantitative Definition of Risk. *Risk Analysis* 1 (1): 11–27. doi:10.1111/j.1539-6924.1981.tb01350.x.

Keck, Margaret E., and Kathryn Sikkink. 1998. *Activists beyond Borders: Advocacy Networks in International Politics*. Ithaca: Cornell University Press.

Keohane, Robert O. 1984. *After Hegemony: Cooperation and Discord in the World Political Economy*. Princeton: Princeton University Press.

Keohane, Robert O., and Joseph S. Nye. 1977. *Power and Interdependence*. Boston: Longman.

Keohane, Robert O., and David G. Victor. 2011. The Regime Complex for Climate Change. *Perspectives on Politics* 9 (1): 7–23. doi:10.1017/S1537592710004068.

Knopf, Brigitte, Martin Kowarsch, Christian Flachsland, and Ottmar Edenhofer. 2012. The 2°C Target Reconsidered. In *Climate Change, Justice and Sustainability*, ed. Ottmar Edenhofer, Johannes Wallacher, Hermann Lotze-Campen, Michael Reder, Brigitte Knopf, and Johannes Müller, 121–137. Dordrecht: Springer Netherlands. http://link.springer.com/chapter/10.1007/978-94-007-4540-7_12.

Kollmuss, Anja, and Julian Agyeman. 2002. Mind the Gap: Why Do People Act Environmentally and What Are the Barriers to Pro-Environmental Behavior? *Environmental Education Research* 8 (3): 239–260. doi:10.1080/13504620220145401.

Koremenos, Barbara, Charles Lipson, and Duncan Snidal, eds. 2003. *The Rational Design of International Institutions*. Cambridge: Cambridge University Press.

Kvaløy, Berit, Henning Finseraas, and Ola Listhaug. 2012. The Public's Concern for Global Warming: A Cross-National Study of 47 Countries. *Journal of Peace Research* 49 (1): 11–22. doi:10.1177/0022343311425841.

Lackey, Robert T. 2007. Science, Scientists, and Policy Advocacy. *Conservation Biology* 21 (1): 12–17. doi:10.1111/j.1523-1739.2006.00639.x.

Lahsen, Myanna. 2010. The Social Status of Climate Change Knowledge: An Editorial Essay. *Wiley Interdisciplinary Reviews: Climate Change* 1 (1). doi:10.1002/wcc.27.

Lakoff, George. 2008. *The Political Mind: Why You Can't Understand 21st-Century American Politics with an 18th-Century Brain*. New York: Viking.

Lazarus, Richard S. 1982. Thoughts on the Relations between Emotion and Cognition. *American Psychologist* 37 (9): 1019–1024. doi:10.1037/0003-066X.37.9.1019.

Leal-Arcas, Rafael. 2011. Top-Down versus Bottom-Up Approaches for Climate Change Negotiations: An Analysis. *IUP Journal of Governance and Public Policy* 6 (4): 7–52.

Lebow, Richard Ned. 2005. Reason, Emotion and Cooperation. *International Politics* 42 (3): 283–313. doi:10.1057/palgrave.ip.8800113.

Leiserowitz, Anthony. 2006. Climate Change Risk Perception and Policy Preferences: The Role of Affect, Imagery, and Values. *Climatic Change* 77 (1): 45–72. doi:10.1007/s10584-006-9059-9.

Lemoine, Derek M., and Christian P. Traeger. 2012. Tipping Points and Ambiguity in the Economics of Climate Change. Working Paper 18230. Cambridge, MA: National Bureau of Economic Research. http://www.nber.org/papers/w18230.

Lenton, Timothy M. 2011a. Beyond 2°C: Redefining Dangerous Climate Change for Physical Systems. *Wiley Interdisciplinary Reviews: Climate Change* 2 (3): 451–461. doi:10.1002/wcc.107.

Lenton, Timothy M. 2011b. Early Warning of Climate Tipping Points. *Nature Climate Change* 1 (4): 201–209. doi:10.1038/nclimate1143.

Lenton, Timothy M. 2012. Arctic Climate Tipping Points. *Ambio* 41 (1): 10–22. doi:10.1007/s13280-011-0221-x.

Lenton, Timothy M. 2014. Game Theory: Tipping Climate Cooperation. *Nature Climate Change* 4 (1): 14–15. doi:10.1038/nclimate2078.

Lenton, Timothy M., and Juan-Carlos Ciscar. 2012. Integrating Tipping Points into Climate Impact Assessments. *Climatic Change* 117 (3): 585–597. doi:10.1007/s10584-012-0572-8.

Lenton, Timothy M., Hermann Held, Elmar Kriegler, Jim W. Hall, Wolfgang Lucht, Stefan Rahmstorf, and Hans Joachim Schellnhuber. 2008. Tipping Elements in the Earth's Climate System. *Proceedings of the National Academy of Sciences of the United States of America* 105 (6): 1786–1793. doi:10.1073/pnas.0705414105.

Lenton, Timothy M., and Hywel T.P. Williams. 2013. On the Origin of Planetary-Scale Tipping Points. *Trends in Ecology & Evolution* 28 (7): 380–382. doi:10.1016/j.tree.2013.06.001.

Levin, Kelly, Benjamin Cashore, Steven Bernstein, and Graham Auld. 2009. Playing It Forward: Path Dependency, Progressive Incrementalism, and the "Super Wicked" Problem of Global Climate Change. In *IOP Conference Series: Earth and Environmental Science* 6 (February): 2002.

Levin, Kelly, Benjamin Cashore, Steven Bernstein, and Graeme Auld. 2012. Overcoming the Tragedy of Super Wicked Problems: Constraining Our Future Selves to

Ameliorate Global Climate Change. *Policy Sciences* 45 (2): 123–152. doi:10.1007/s11077-012-9151-0.

Levy, David L., and Peter J. Newell. 2004. *The Business of Global Environmental Governance*. Cambridge, MA: MIT Press.

Lezak, Stephen B., and Paul H. Thibodeau. 2016. Systems Thinking and Environmental Concern. *Journal of Environmental Psychology* 46 (June): 143–153. doi:10.1016/j.jenvp.2016.04.005.

Liberman, Nira, and Yaacov Trope. 2008. The Psychology of Transcending the Here and Now. *Science* 322 (5905): 1201–1205. doi:10.1126/science.1161958.

Lindkvist, Emilie, and Jon Norberg. 2014. Modeling Experiential Learning: The Challenges Posed by Threshold Dynamics for Sustainable Renewable Resource Management. *Ecological Economics* 104 (August): 107–118. doi:10.1016/j.ecolecon.2014.04.018.

Linklater, Andrew. 2014. Anger and World Politics: How Collective Emotions Shift over Time. *International Theory* 6 (3): 574–578. doi:10.1017/S1752971914000293.

Liverman, Diana M. 2009. Conventions of Climate Change: Constructions of Danger and the Dispossession of the Atmosphere. *Journal of Historical Geography* 35 (2): 279–296. doi:10.1016/j.jhg.2008.08.008.

Loewenstein, George. 1992. *Choice over Time*. Ed. Jon Elster. New York: Russell Sage Foundation.

Loewenstein, George F., Elke U. Weber, Christopher K. Hsee, and Ned Welch. 2001. Risk as Feelings. *Psychological Bulletin* 127 (2): 267–286. doi:10.1037/0033-2909.127.2.267.

Lorenzoni, Irene, Anthony Leiserowitz, De Franca Doria Miguel, Wouter Poortinga, and Nick F. Pidgeon. 2006. Cross-National Comparisons of Image Associations with "Global Warming" and "Climate Change" among Laypeople in the United States of America and Great Britain. *Journal of Risk Research* 9 (3): 265. doi:10.1080/13669870600613658.

Lowe, Thomas D., and Irene Lorenzoni. 2007. Danger Is All around: Eliciting Expert Perceptions for Managing Climate Change through a Mental Models Approach. *Global Environmental Change* 17 (1): 131–146. doi:10.1016/j.gloenvcha.2006.05.001.

Lustick, Ian S., and Dan Miodownik. 2009. Abstractions, Ensembles, and Virtualizations: Simplicity and Complexity in Agent-Based Modeling. *Comparative Politics* 41 (January): 223–244. doi:10.5129/001041509X12911362972070.

Maguire, Steve. 2004. The Co-Evolution of Technology and Discourse: A Study of Substitution Processes for the Insecticide DDT. *Organization Studies* 25 (1): 113–134. doi:10.1177/0170840604038183.

Maguire, Steve, and Cynthia Hardy. 2009. Discourse and Deinstitutionalization: The Decline of DDT. *Academy of Management Journal ARCHIVE* 52 (1): 148–178.

Marcus, George E. 2000. Emotions in Politics. *Annual Review of Political Science* 3 (1): 221–250. doi:10.1146/annurev.polisci.3.1.221.

Marcus, George E., W. Russell Neuman, Michael MacKuen, and Ann N. Crigler. 2007. *The Affect Effect: Dynamics of Emotion in Political Thinking and Behavior.* Chicago: University of Chicago Press.

Markowitz, Ezra M. 2012. Is Climate Change an Ethical Issue? Examining Young Adults' Beliefs about Climate and Morality. *Climatic Change* 114 (3–4): 479–495. doi:10.1007/s10584-012-0422-8.

Markowitz, Ezra M., and Azim F. Shariff. 2012. Climate Change and Moral Judgement. *Nature Climate Change* 2 (4): 243–247. doi:10.1038/nclimate1378.

Markus, Hazel R. and Maryam G. Hamedani. 2010. Sociocultural Psychology. In *Handbook of Cultural Psychology,* ed. Shinobu Kitayama and Dov Cohen, 3–39. New York: Guilford Press.

Matthews, Paul. 2015. Why Are People Skeptical about Climate Change? Some Insights from Blog Comments. *Environmental Communication* 9 (2): 153–168. doi:10.1080/17524032.2014.999694.

McCright, Aaron M., and Riley E. Dunlap. 2000. Challenging Global Warming as a Social Problem: An Analysis of the Conservative Movement's Counter-Claims. *Social Problems* 47 (4): 499–522.

McDermott, Rose. 2004. The Feeling of Rationality: The Meaning of Neuroscientific Advances for Political Science. *Perspectives on Politics* 2 (4): 691–706. doi:10.1017/S1537592704040459.

McDermott, Rose. 2014. The Body Doesn't Lie: A Somatic Approach to the Study of Emotions in World Politics. *International Theory* 6 (3): 557–562. doi:10.1017/S1752971914000268.

McDoom, Omar Shahabudin. 2012. The Psychology of Threat in Intergroup Conflict: Emotions, Rationality, and Opportunity in the Rwandan Genocide. *International Security* 37 (2): 119–155.

McGeer, Victoria. 2004. The Art of Good Hope. *Annals of the American Academy of Political and Social Science* 592 (1): 100–127. doi:10.1177/0002716203261781.

Mercer, Jonathan. 2005. Rationality and Psychology in International Politics. *International Organization* 59 (1): 77–106. doi:10.1017/S0020818305050058.

Mercer, Jonathan. 2010. Emotional Beliefs. *International Organization* 64 (1): 1–31. doi:10.1017/S0020818309990221.

Mercer, Jonathan. 2014. Feeling like a State: Social Emotion and Identity. *International Theory* 6 (3): 515–535. doi:10.1017/S1752971914000244.

Merton, Robert K. 1940. Bureaucratic structure and personality. *Social Forces* 18 (4): 560–568.

Meyer, Morgan. 2010. The Rise of the Knowledge Broker. *Science Communication* 32 (1): 118–127. doi:10.1177/1075547009359797.

Michaelowa, Katharina, and Axel Michaelowa. 2012. Negotiating Climate Change. *Climate Policy* 12 (5): 527–533. doi:10.1080/14693062.2012.693393.

Milkoreit, Manjana. 2014. Hot Deontology and Cold Consequentialism—an Empirical Exploration of Ethical Reasoning among Climate Change Negotiators. *Climatic Change* 130 (3): 1–13. doi:10.1007/s10584-014-1170-8.

Milkoreit, Manjana. 2015. Science and Climate Change Diplomacy: Cognitive Limits and the Need to Reinvent Science Communication. In *Science Diplomacy: New Day or False Dawn*, ed. Lloyd Spencer Davis and Robert G. Patman, 109–131. Hackensack, NJ: World Scientific. http://www.worldscientific.com/doi/abs/10.1142/9789814440073_0006.

Milkoreit, Manjana. 2016. The Promise of Climate Fiction: Imagination, Storytelling, and the Politics of the Future. In *Reimagining Climate Change*, ed. Paul Wapner and Hilal Elver, 71–91. Routledge Advances in Climate Change Research. London: Routledge, Taylor & Francis Group.

Milkoreit, Manjana, and Steven Mock. 2014. The Networked Mind: Collective Identities and the Cognitive-Affective Nature of Conflict. In *Networks and Network Analysis for Defence and Security*, ed. Anthony J. Masys, 161–188. Lecture Notes in Social Networks. Cham, Switzerland: Springer International. http://link.springer.com/chapter/10.1007/978-3-319-04147-6_7.

Miller, Clark. 2001. Hybrid Management: Boundary Organizations, Science Policy, and Environmental Governance in the Climate Regime. *Science, Technology & Human Values* 26 (4): 478–500. doi:10.1177/016224390102600405.

Miller, Clark, and Paul N. Edwards. 2001. *Changing the Atmosphere: Expert Knowledge and Environmental Governance*. Cambridge, MA: MIT Press.

Mills, Charles Wright. 1959. *The Sociological Imagination*. New York: Oxford University Press.

Mitchell, Ronald B. 2006. Problem Structure, Institutional Design, and the Relative Effectiveness of International Environmental Agreements. *Global Environmental Politics* 6 (3): 72–89. doi:10.1162/glep.2006.6.3.72.

Moïsi, Dominique. 2009. *The Geopolitics of Emotion: How Cultures of Fear, Humiliation, and Hope Are Reshaping the World*. New York: Doubleday.

Moser, Susanne C. 2010. Communicating Climate Change: History, Challenges, Process and Future Directions. *Wiley Interdisciplinary Reviews: Climate Change* 1 (1): 31–53. doi:10.1002/wcc.11.

Moser, Susanne C., and Lisa Dilling. 2007. *Creating a Climate for Change: Communicating Climate Change and Facilitating Social Change.* Cambridge: Cambridge University Press.

Moten, David, and Simon Niemeyer. 2008. *AdvanvceQ (Java Based Computer Program).* Canberra: Australian National University.

Myers, Teresa A., Edward W. Maibach, Connie Roser-Renouf, Karen Akerlof, and Anthony A. Leiserowitz. 2013. The Relationship between Personal Experience and Belief in the Reality of Global Warming. *Nature Climate Change* 3 (4): 343–347. doi:10.1038/nclimate1754.

Najam, Adil. 2005. Developing Countries and Global Environmental Governance: From Contestation to Participation to Engagement. *International Environmental Agreement: Politics, Law and Economics* 5 (3): 303–321. doi:10.1007/s10784-005 -3807-6.

Nelson, Thomas E. 1997. Toward a Psychology of Framing Effects. *Political Behavior* 19 (3): 221–246.

Niemeyer, Simon. 2004. Deliberation in the Wilderness: Displacing Symbolic Politics. *Environmental Politics* 13 (2): 347–372.

Niemeyer, Simon. 2011. The Emancipatory Effect of Deliberation: Empirical Lessons from Mini-Publics. *Politics & Society* 39 (1): 103–140. doi:10.1177/0032329210395000.

Nisbett, Richard E., and Lee Ross. 1980. *Human Inference: Strategies and Shortcomings of Social Judgment.* Englewood Cliffs, NJ: Prentice-Hall.

Nordhaus, William D. 2012. Economic Policy in the Face of Severe Tail Events. *Journal of Public Economic Theory* 14 (2): 197–219. doi:10.1111/j.1467-9779.2011.01544.x.

Norgaard, Kari Marie. 2006a. People Want to Protect Themselves a Little Bit: Emotions, Denial, and Social Movement Nonparticipation. *Sociological Inquiry* 76 (August): 372–396. doi:10.1111/j.1475-682X.2006.00160.x.

Norgaard, Kari Marie. 2006b. "We Don't Really Want to Know": Environmental Justice and Socially Organized Denial of Global Warming in Norway. *Organization & Environment* 19 (3): 347–370. doi:10.1177/1086026606292571.

Norgaard, Kari Marie. 2009. Cognitive and Behavioral Challenges in Responding to Climate Change. *SSRN eLibrary*, May. http://papers.ssrn.com/sol3/papers.cfm ?abstract_id=1407958.

Norgaard, Kari Marie. 2011. *Living in Denial: Climate Change, Emotions, and Everyday Life.* Cambridge, MA: MIT Press.

Novak, Joseph D. 1998. *Learning, Creating, and Using Knowledge: Concept Maps as Facilitative Tools in Schools and Corporations.* New York: Routledge.

Nuttall, Mark. 2012. Tipping Points and the Human World: Living with Change and Thinking about the Future. *Ambio* 41 (1): 96–105. doi:10.1007/s13280-011-0228-3.

Okereke, Chukwumerije. 2010. Climate Justice and the International Regime. *Wiley Interdisciplinary Reviews: Climate Change* 1 (3): 462–474. doi:10.1002/wcc.52.

O'Neill, Saffron J., Maxwell Boykoff, Simon Niemeyer, and Sophie A. Day. 2013. On the Use of Imagery for Climate Change Engagement. *Global Environmental Change* 23 (2): 413–421. doi:10.1016/j.gloenvcha.2012.11.006.

Oreskes, Naomi, and Erik M. Conway. 2011. *Merchants of Doubt: How a Handful of Scientists Obscured the Truth on Issues from Tobacco Smoke to Global Warming.* New York: Bloomsbury Press.

Ostrom, Elinor. 2010. Polycentric Systems for Coping with Collective Action and Global Environmental Change. *Global Environmental Change* 20 (4): 550–557.

Page, Edward. 2013. Climate Change Justice. In *Handbook of Global Climate and Environmental Policy*, ed. Robert Falkner, 231–247. Chichester, England: Wiley-Blackwell.

Paterson, Matthew. 2009. Post-Hegemonic Climate Politics? *British Journal of Politics and International Relations* 11 (1): 140–158. doi:10.1111/j.1467-856X.2008.00354.x.

Pattberg, Philipp, and Johannes Stripple. 2008. Beyond the Public and Private Divide: Remapping Transnational Climate Governance in the 21st Century. *International Environmental Agreement: Politics, Law and Economics* 8 (4): 367–388. doi:10.1007/s10784-008-9085-3.

Pattberg, Philipp, and Oscar Widerberg. 2015. Theorising Global Environmental Governance: Key Findings and Future Questions. *Millennium* 43 (2): 684–705. doi:10.1177/0305829814561773.

Pessoa, Luiz. 2008. On the Relationship between Emotion and Cognition. *Nature Reviews Neuroscience* 9 (2): 148–158. doi:10.1038/nrn2317.

Peters, Glen P., Robbie M. Andrew, Tom Boden, Josep G. Canadell, Philippe Ciais, Corinne Le Quéré, et al. 2013. The Challenge to Keep Global Warming below 2°C. *Nature Climate Change* 3 (1): 4–6. doi:10.1038/nclimate1783.

Pettenger, Mary E. 2007. *The Social Construction of Climate Change: Power, Knowledge, Norms, Discourses.* Aldershot, England: Ashgate.

Phillips, Leigh. 2012. Sea versus Senators. *Nature* 486 (7404): 450. doi:10.1038/486450a.

Pickering, Jonathan, Steve Vanderheiden, and Seumas Miller. 2012. "If Equity's In, We're Out": Scope for Fairness in the Next Global Climate Agreement. *Ethics & International Affairs* 26 (4): 423–443.

Prins, Gwyn, Isabel Galiana, Christopher Green, Reiner Grundmann, Atte Korhola, Frank Laird, et al. 2010. The Hartwell Paper: A New Direction for Climate Policy after the Crash of 2009. Monograph. April. http://eprints.lse.ac.uk/27939/.

Proshansky, Harold M. 1978. The City and Self-Identity. *Environment and Behavior* 10 (2): 147–169. doi:10.1177/0013916578102002.

Proshansky, Harold M., Abbe K. Fabian, and Robert Kaminoff. 1983. Place-Identity: Physical World Socialization of the Self. *Journal of Environmental Psychology* 3 (1): 57–83. doi:10.1016/S0272-4944(83)80021-8.

Putnam, Robert D. 1988. Diplomacy and Domestic Politics: The Logic of Two-Level Games. *International Organization* 42 (3): 427–460. doi:10.1017/S0020818300027697.

Randalls, Samuel. 2010. History of the 2°C Climate Target. *Wiley Interdisciplinary Reviews: Climate Change* 1 (4): 598–605. doi:10.1002/wcc.62.

Rathwell, Kaitlyn Joanne, Derek Armitage, and Fikret Berkes. 2015. Bridging Knowledge Systems to Enhance Governance of the Environmental Commons: A Typology of Settings. *International Journal of the Commons* 9 (2): 851–880.

Rayner, Steve. 1991. A Cultural Perspective on the Structure and Implementation of Global Environmental Agreements. *Evaluation Review* 15 (1): 75–102. doi:10.1177/01 93841X9101500105.

Renshon, Jonathan. 2008. Stability and Change in Belief Systems: The Operational Code of George W. Bush. *Journal of Conflict Resolution* 52 (6): 820–849. doi:10.1177/ 0022002708323669.

Reser, Joseph P., Graham L. Bradley, and Michelle C. Ellul. 2014. Encountering Climate Change: "Seeing" Is More than "Believing." *Wiley Interdisciplinary Reviews: Climate Change* 5 (4): 521–537. doi:10.1002/wcc.286.

Reus-Smit, Christian. 2004. *The Politics of International Law.* Cambridge University Press.

Reus-Smit, Christian. 2014. Emotions and the Social. *International Theory* 6 (3): 568–574. doi:10.1017/S1752971914000281.

Ringius, Lasse, Asbjørn Torvanger, and Arild Underdal. 2002. Burden Sharing and Fairness Principles in International Climate Policy. *International Environmental Agreement: Politics, Law and Economics* 2 (1): 1–22.

Risse, Thomas, Daniela Engelmann-Martin, Hans-Joachim Knope, and Klaus Roscher. 1999. To Euro or Not to Euro? The EMU and Identity Politics in the European Union. *European Journal of International Relations* 5 (2): 147–187. doi:10.1177/1 354066199005002001.

Rittel, Horst W. J., and Melvin M. Webber. 1973. Dilemmas in a General Theory of Planning. *Policy Sciences* 4 (2): 155–169. doi:10.1007/BF01405730.

Roberts, J. Timmons, and Bradley C. Parks. 2006. *A Climate of Injustice: Global Inequality, North-South Politics, and Climate Policy.* Cambridge, MA: MIT Press.

Roeser, Sabine. 2010. Intuitions, Emotions and Gut Reactions in Decisions about Risks: Towards a Different Interpretation of "Neuroethics." *Journal of Risk Research* 13 (2): 175–190. doi:10.1080/13669870903126275.

Roeser, Sabine. 2011. *Moral Emotions and Intuitions.* Basingstroke, England: Palgrave Macmillan.

Roeser, Sabine. 2012. Risk Communication, Public Engagement, and Climate Change: A Role for Emotions. *Risk Analysis* 32 (6): 1033–1040. doi:10.1111/j.1539-6924.2012.01812.x.

Rogelj, Joeri, William Hare, Jason Lowe, Detlef P. van Vuuren, Keywan Riahi, Tatsuya Hanaoka Ben Matthews, et al. 2011. Emission Pathways Consistent with a 2°C Global Temperature Limit. *Nature Climate Change* 1 (8): 413–418. doi:10.1038/nclimate1258.

Rogelj, Joeri, David L. McCollum, Brian C. O'Neill, and Keywan Riahi. 2011. 2020 Emissions Levels Required to Limit Warming to below 2°C. *Nature Climate Change* 3 (4): 405–412. doi:10.1038/nclimate1758.

Rogelj, Joeri, David L. McCollum, Andy Reisinger, Malte Meinshausen, and Keywan Riahi. 2013. Probabilistic Cost Estimates for Climate Change Mitigation. *Nature* 493 (7430): 79–83. doi:10.1038/nature11787.

Roser-Renouf, Connie, Edward W. Maibach, Anthony Leiserowitz, and Xiaoquan Zhao. 2014. The Genesis of Climate Change Activism: From Key Beliefs to Political Action. *Climatic Change* 125 (2): 1–16. doi:10.1007/s10584-014-1173-5.

Roser-Renouf, Connie, Neil Stenhouse, Justin Rolfe-Redding, Edward W. Maibach, and Anthony Leiserowitz. 2015. Engaging Diverse Audiences with Climate Change: Message Strategies for Global Warming's Six Americas. In *The Routledge Handbook of Environment and Communication,* ed. Anders Hansen and Robert Cox, 368–386. London: Routledge. http://papers.ssrn.com/sol3/papers.cfm?abstract_id=2410650.

Sacchi, Simona, Paolo Riva, Marco Brambilla, and Marco Grasso. 2014. Moral Reasoning and Climate Change Mitigation: The Deontological Reaction toward the Market-Based Approach. *Journal of Environmental Psychology* 38 (June): 252–261. doi:10.1016/j.jenvp.2014.03.001.

Safi, Ahmad Saleh, William James Smith Jr., and Zhnongwei Liu. 2012. Rural Nevada and Climate Change: Vulnerability, Beliefs, and Risk Perception. *Risk Analysis* 32 (6): 1041–1059. doi:10.1111/j.1539-6924.2012.01836.x.

Sandbrook, Chris, Fred Nelson, William M. Adams, and Arun Agrawal. 2010. Carbon, Forests and the REDD Paradox. *Oryx* 44 (3): 330–334. doi:10.1017/S0030605310000475.

Sasley, Brent E. 2011. Theorizing States' Emotions. *International Studies Review* 13 (3): 452–476. doi:10.1111/j.1468-2486.2011.01049.x.

Schelling, Thomas C. 2000. Intergenerational and International Discounting. *Risk Analysis* 20 (6): 833–838. doi:10.1111/0272-4332.206076.

Scholl, Wolfgang. 2013. The Socio-Emotional Basis of Human Interaction and Communication: How We Construct Our Social World. *Social Sciences Information/ Information sur les Sciences Sociales* 52 (1): 3–33. doi:10.1177/0539018412466607.

Schroeder, Heike, and Heather Lovell. 2012. The Role of Non-Nation-State Actors and Side Events in the International Climate Negotiations. *Climate Policy* 12 (1): 23–37. doi:10.1080/14693062.2011.579328.

Serdeczny, Olivia, Eleanor Waters, and Sander Chan. 2016. Non-Economic Loss and Damage in the Context of Climate Change: Understanding the Challenges. Bonn: German Development Institute. https://www.die-gdi.de/en/discussion-paper/article/non-economic-loss-and-damage-in-the-context-of-climate-change-understanding-the-challenges/.

Shapiro, Michael J., G. Matthew Bonham, and Daniel Heradstveit. 1988. A Discursive Practices Approach to Collective Decision-Making. *International Studies Quarterly* 32 (4): 397–419. doi:10.2307/2600590.

Sheppard, Stephen R. J., Alison Shaw, David Flanders, Sarah Burch, Arnim Wiek, Jeff Carmichael, et al. 2011. Future Visioning of Local Climate Change: A Framework for Community Engagement and Planning with Scenarios and Visualisation. *Futures* 43 (4): 400–412. Special Issue: *Community Engagement for Sustainable Urban Futures.* doi:10.1016/j.futures.2011.01.009.

Sjöberg, Lennart. 1998. Worry and Risk Perception. *Risk Analysis* 18 (1): 85–93. doi:10.1111/j.1539-6924.1998.tb00918.x.

Sjöberg, Lennart. 2000. Factors in Risk Perception. *Risk Analysis* 20 (1): 1–12. doi:10.1111/0272-4332.00001.

Skitka, Linda J., Christopher W. Bauman, and Edward G. Sargis. 2005. Moral Conviction: Another Contributor to Attitude Strength or Something More? *Journal of Personality and Social Psychology* 88 (6): 895–917. doi:10.1037/0022-3514.88.6.895.

Slovic, Paul E. 2000. *The Perception of Risk.* London: Earthscan Publications.

Slovic, Paul, and Elke Weber. 2002. Perception of Risk Posed by Extreme Events. Center for Decision Science (CDS) Working Paper. New York: Columbia University. https://www.ldeo.columbia.edu/chrr/documents/meetings/roundtable/white _papers/slovic_wp.pdf.

Smith, Nicholas, and Anthony Leiserowitz. 2012. The Rise of Global Warming Skepticism: Exploring Affective Image Associations in the United States over Time. *Risk Analysis* 32 (6): 1021–1032. doi:10.1111/j.1539-6924.2012.01801.x.

Smith, Tammy. 2007. Narrative Boundaries and the Dynamics of Ethnic Conflict and Conciliation. *Poetics* 35 (1): 22–46. doi:10.1016/j.poetic.2006.11.001.

Snyder, Charles R. 2002. Hope Theory: Rainbows in the Mind. *Psychological Inquiry* 13 (4): 249–275.

Spence, Alexa, Wouter Poortinga, Catherine Butler, and Nick F. Pidgeon. 2011. Perceptions of Climate Change and Willingness to Save Energy Related to Flood Experience. *Nature Climate Change* 1 (1): 46–49. doi:10.1038/nclimate1059.

Spence, Alexa, Wouter Poortinga, and Nick Pidgeon. 2012. The Psychological Distance of Climate Change. *Risk Analysis* 32 (6): 957–972. doi:10.1111/ j.1539-6924.2011.01695.x.

Staiger, Janet, Ann Cvetkovich, and Ann Reynolds. 2010. *Political Emotions.* eBook. Hoboken, NJ: Taylor & Francis.

Stephenson, William. 1935. Technique of Factor Analysis. *Nature* 136 (297). doi:10.1038/136297b0.

Stephenson, William. 1953. *The Study of Behavior: Q-Technique and Its Methodology.* Chicago: University of Chicago Press.

Stern, Nicholas. 2007. *The Economics of Climate Change: The Stern Review.* Cambridge: Cambridge University Press.

Stern, Paul C. 2012. Fear and Hope in Climate Messages. *Nature Climate Change* 2 (8): 572–573. doi:10.1038/nclimate1610.

Stocker, Thomas F. 2013. The Closing Door of Climate Targets. *Science* 339 (6117): 280–282. doi:10.1126/science.1232468.

Strange, Susan. 1996. *The Retreat of the State: The Diffusion of Power in the World Economy.* Cambridge: Cambridge University Press.

Sunstein, Cass R. 2007. Of Montreal and Kyoto: A Tale of Two Protocols. *Harvard Environmental Law Review* 31:1.

Tajfel, Henri. 1982. *Social Identity and Intergroup Relations.* Cambridge: Cambridge University Press.

Tanner, Carmen, Douglas L. Medin, and Rumen Iliev. 2008. Influence of Deontological versus Consequentialist Orientations on Act Choices and Framing Effects: When Principles Are More Important than Consequences. *European Journal of Social Psychology* 38 (5): 757–769. doi:10.1002/ejsp.493.

Thagard, Paul. 2000. *Coherence in Thought and Action. Life and Mind*. Cambridge, MA: MIT Press.

Thagard, Paul. 2005. *Mind: Introduction to Cognitive Science*. Rev. ed. Cambridge, MA: MIT Press.

Thagard, Paul. 2006. *Hot Thought: Mechanisms and Applications of Emotional Cognition*. Cambridge, MA: MIT Press.

Thagard, Paul. 2008. How Cognition Meets Emotion: Beliefs, Desires and Feelings and Neural Activity. In *Epistemology and Emotions*, ed. Georg Brun, Ulvi Doguoglu, and Dominique Kuenzle, 167–184. Aldershot, England: Ashgate.

Thagard, Paul. 2010a. EMPATHICA: A Computer Support System with Visual Representations for Cognitive-Affective Mapping. In *Proceedings of the Workshop on Visual Reasoning and Representation*, 79–81. Menlo Park, CA: AAAI Press.

Thagard, Paul. 2010b. *The Brain and the Meaning of Life*. Princeton: Princeton University Press.

Thagard, Paul. 2010c. Explaining Economic Crises: Are There Collective Representations? *Episteme* 7 (3): 266–283. doi:10.3366/epi.2010.0207.

Thagard, Paul. 2012. Mapping Minds across Cultures. In *Grounding Social Sciences in Cognitive Sciences*, ed. Ron Sun, 35–62. Cambridge, MA: MIT Press.

Tol, Richard. 2007. Europe's Long-Term Climate Target: A Critical Evaluation. *Energy Policy* 35 (1): 424–432. doi:10.1016/j.enpol.2005.12.003.

Tonn, Bruce, Angela Hemrick, and Fred Conrad. 2006. Cognitive Representations of the Future: Survey Results. *Futures* 38 (7): 810–829. doi:10.1016/j.futures.2005.12.005.

Torcello, Lawrence. 2016. The Ethics of Belief, Cognition, and Climate Change Pseudoskepticism: Implications for Public Discourse. *Topics in Cognitive Science* 8 (1): 19–48. doi:10.1111/tops.12179.

Tschakert, Petra, and Kathleen Ann Dietrich. 2010. Anticipatory Learning for Climate Change Adaptation and Resilience. *Ecology and Society* 15 (2): 11.

Turner, Mark. 2003. *Cognitive Dimensions of Social Science: The Way We Think about Politics, Economics, Law, and Society*. Oxford: Oxford University Press.

Underdal, Arild. 1992. The Concept of Regime "Effectiveness." *Cooperation and Conflict* 27 (3): 227–240. doi:10.1177/0010836792027003001.

Underdal, Arild. 2010. Complexity and Challenges of Long-Term Environmental Governance. *Global Environmental Change* 20 (3): 386–393. doi:10.1016/j. gloenvcha.2010.02.005.

Verweij, Marco, Mary Douglas, Richard Ellis, Christoph Engel, Frank Hendriks, Susanne Lohmann, et al. 2006. Clumsy Solutions for a Complex World: The Case of Climate Change. *Public Administration* 84 (4): 817–843. doi:10.1111/ j.1540-8159.2005.09566.x-i1.

Victor, David G. 2006. Toward Effective International Cooperation on Climate Change: Numbers, Interests and Institutions. *Global Environmental Politics* 6 (3): 90–103.

Victor, David G. 2011. *Global Warming Gridlock: Creating More Effective Strategies for Protecting the Planet.* Cambridge: Cambridge University Press.

Vohs, Kathleen D., Roy F. Baumeister, and George Loewenstein. 2007. *Do Emotions Help or Hurt Decision Making? A Hedgefoxian Perspective.* New York: Russell Sage Foundation.

Wagner, Gernot. 2012. Climate Policy: Hard Problem, Soft Thinking. *Climatic Change* 110 (3–4): 507–521.

Wang, Chi-Hsiang, and Jane M. Blackmore. 2009. Resilience Concepts for Water Resource Systems. *Journal of Water Resources Planning and Management* 135 (6): 528–536. doi:10.1061/(ASCE)0733-9496(2009)135:6(528).

Ward, Hugh. 1996. Game Theory and the Politics of Global Warming: The State of Play and Beyond. *Political Studies* 44 (5): 850–871. doi:10.1111/j.1467-9248.1996. tb00338.x.

Watts, Simon, and Paul Stenner. 2005. Doing Q Methodology: Theory, Method and Interpretation. *Qualitative Research in Psychology* 2 (1): 67–91. doi:10.1191/14780887 05qp022oa.

Watts, Simon, and Paul Stenner. 2012. *Doing Q Methodological Research: Theory, Method, and Interpretation.* Los Angeles: Sage.

Weart, Spencer R. 2010. The Idea of Anthropogenic Global Climate Change in the 20th Century. *Wiley Interdisciplinary Reviews: Climate Change* 1 (1): 67–81. doi:10. 1002/wcc.6.

Weber, Elke U. 2010. What Shapes Perceptions of Climate Change? *Wiley Interdisciplinary Reviews: Climate Change* 1 (3): 332–342. doi:10.1002/wcc.41.

Weintrobe, Sally, ed. 2012. *Engaging with Climate Change: Psychoanalytic and Interdisciplinary Perspectives.* London: Routledge.

Weischer, Lutz, Jennifer Morgan, and Milap Patel. 2012. Climate Clubs: Can Small Groups of Countries Make a Big Difference in Addressing Climate Change? *Review of*

European Community & International Environmental Law 21 (3): 177–192. doi:10.1111/reel.12007.

Weitzman, Martin L. 2009. On Modeling and Interpreting the Economics of Catastrophic Climate Change. *Review of Economics and Statistics* 91 (1): 1–19. doi:10.1162/rest.91.1.1.

Wendt, Alexander. 1992. Anarchy Is What States Make of It: The Social Construction of Power Politics. *International Organization* 46 (2): 391–425.

Wendt, Alexander. 1999. *Social Theory of International Politics*. Cambridge: Cambridge University Press.

Werrell, Caitlin E., Francesco Femia, and Anne-Marie Slaughter. 2012. The Arab Spring and Climate Change. Washington, DC: Center for American Progress. https://www.americanprogress.org/issues/security/report/2013/02/28/54579/the-arab-spring-and-climate-change/.

Westen, Drew. 2008. *The Political Brain: The Role of Emotion in Deciding the Fate of the Nation*. New York: PublicAffairs.

Westley, Frances, Per Olsson, Carl Folke, Thomas Homer-Dixon, Harrie Vredenburg, Derk Loorbach, et al. 2011. Tipping toward Sustainability: Emerging Pathways of Transformation. *Ambio* 40 (7): 762–780. doi:10.1007/s13280-011-0186-9.

Wiek, Arnim, and David Iwaniec. 2013. Quality Criteria for Visions and Visioning in Sustainability Science. *Sustainability Science* 9 (4): 1–16. doi:10.1007/s11625-013-0208-6.

Wolf, Johanna, and Susanne C. Moser. 2011. Individual Understandings, Perceptions, and Engagement with Climate Change: Insights from In-depth Studies across the World. *Wiley Interdisciplinary Reviews: Climate Change* 2 (4): 547–569. doi:10.1002/wcc.120.

Young, Michael D., and Mark Schafer. 1998. Is There Method in Our Madness? Ways of Assessing Cognition in International Relations. *Mershon International Studies Review* 42 (1): 63–96. doi:10.2307/254444.

Young, Oran R. 1994. *International Governance: Protecting the Environment in a Stateless Society*. Ithaca: Cornell University Press.

Young, Oran R. 1997. Rights, Rules and Resources in World Affairs. In *Global Governance: Drawing Insights from the Environmental Experience*, ed. Oran R. Young, 1–24. Cambridge, MA: MIT Press.

Young, Oran R. 2011. Effectiveness of International Environmental Regimes: Existing Knowledge, Cutting-Edge Themes, and Research Strategies. *Proceedings of the National Academy of Sciences of the United States of America* 108 (50): 19853–19860. doi:10.1073/pnas.1111690108.

Yusoff, Kathryn, and Jennifer Gabrys. 2011. Climate Change and the Imagination. *Wiley Interdisciplinary Reviews: Climate Change* 2 (4): 516–534. doi:10.1002/wcc.117.

Zajonc, R. B. 1980. Feeling and Thinking: Preferences Need No Inferences. *American Psychologist* 35 (2): 151–175.

Zaval, Lisa, Elizabeth A. Keenan, Eric J. Johnson, and Elke U. Weber. 2014. How Warm Days Increase Belief in Global Warming. *Nature Climate Change* 4 (2): 143–147.

Zelli, Fariborz. 2011. The Fragmentation of the Global Climate Governance Architecture. *Wiley Interdisciplinary Reviews: Climate Change* 2 (2): 255–270. doi:10.1002/wcc.104.

Zia, Asim. 2013. *Post-Kyoto Climate Governance: Confronting the Politics of Scale, Ideology and Knowledge.* New York: Routledge.

Index